W9-BRL-058

Wires West

BOOKS BY PHIL AULT

WIRES WEST

THESE ARE THE GREAT LAKES

THIS IS THE DESERT

WONDERS OF THE MOSQUITO WORLD

WIRES WEST

Phil Ault

ILLUSTRATED WITH PHOTOGRAPHS,

OLD PRINTS, AND MAPS

Dodd, Mead & Company, New York

Illustrations courtesy of:

American Telephone and Telegraph Company, 92, 147, 149, 150, 152, 153, 154, 156 *bottom,* 158 *top,* 159, 161, 162; Archives of Reverend Eugene Buechel, S.J., St. Francis Mission, St. Francis, South Dakota, 170-171; Creighton University, 42 *bottom,* 59; The Growth of Industrial Art, Government Printing Office, 151, 165; *Harper's Monthly,* 12, 139, 156 *top; Harper's Weekly,* 3, 7, 31, 35, 87, 98, 116, 137, 143; *Harper's Weekly,* from the Library of Congress, 142; *Leslie's Weekly,* 46; *Leslie's Magazine,* from the Library of Congress, 122, 127; Library of Congress, 5, 40, 54, 130; Treasure Products, Inc., publisher of the C.M. Russell *Boyhood Sketchbook,* Bozeman, Montana, 63, 65; Union Pacific Railroad Photo, 10, 49, 68, 94, 119, 121, 125; United States Department of the Interior, National Park Service, photo by William S. Keller, 8 *bottom;* Utah State Historical Society, 74; Western Union Photo, viii, 17, 20, 21, 33, 47, 61, 75, 134, 135, 158 *bottom,* 167, 168.

Maps on pages 38, 56, 84, and 102-103 by Salem Tamer

ISBN: 0-396-06968-1
Library of Congress Catalog Card Number; 74-2603
Printed in the United States of America

For Henry, Ben, Sam, Walter and
the other Morse operators I worked beside,
a long time ago

Contents

1. Across the Grassy Sea 1
2. The Telegraph Is Born 11
3. Gold, Ponies, and Wires 27
4. Stringing the Wire 41
5. Charley Brown's Diary 52
6. Hello, San Francisco 64
7. The Indians Attack 77
8. The Wire Talks Again 89
9. On to Russia 96
10. "The Craziest Thing Ever" 111
11. Here Come the Trains 118
12. Men in Green Eyeshades 133
13. "Mr. Watson, Come Here" 146
14. Farewell, Brass Pounders 164
Acknowledgments and Bibliography 173
Index 175

Samuel F. B. Morse as he appeared when he invented the first practical telegraph

CHAPTER 1

Across the Grassy Sea

A tiny dark dot came into sight on the western skyline, like a distant ship breaking the horizon of a gently rolling sea. But this was an ocean of shaggy grass, dry and sparse and seemingly endless. Swept along by the west wind, the dot came closer, trailing dust behind it as a ship leaves a wake. Not a tree was visible to break the emptiness of the prairie; nothing but land and sky, and the dot growing larger.

The men standing outside a sod hut soon were satisfied that it was what they had been waiting for—a man on horseback.

"He's coming, finally," one said in a tone of relief. "Late, though. Should have been here nearly three hours ago."

The other steadied the horse he held, turning its head to the east—a fresh horse saddled and ready, lean and long legged; a little wild perhaps but chosen for its speed and durability. "He'll not be wasting many seconds here."

Galloping closer along the dirt track toward the solitary building, the horse and rider emerged as distinct figures, both sweaty from the high-speed run. Half a mile away the rider waved his hat and let forth a high, mournful yell like a coyote's to announce his arrival; his way of saying, "Get ready for me."

At the station—for that's what it was, a station of the Pony Express—the rider reined in his horse and jumped to the ground. Breathlessly he acknowledged the "Hello, Jim" from the man holding the fresh horse. The mail was late, no time even for the usual pleasantries.

"Trouble on the line," Jim Moore panted. "Indians. Half a dozen of them on the trail. I had to cut across country to get away."

Without more talk, Jim seized the *mochila* from the saddle of his panting horse and tossed it across the back of the fresh one. The *mochila* was a small leather sheet whose ends fell on either side of the horse, with two pockets or *cantinas* on each side fore and aft of the rider's leg. Inside the *cantinas,* wrapped in oil silk, was the precious Pony Express cargo, packets of letters from San Francisco being carried across nearly two thousand miles of wilderness through mountains and across the prairies to St. Joseph, Missouri. That little waterside settlement on the Missouri River was for all purposes the western outpost of American civilization on this summer day in 1861.

Moore climbed briskly into the saddle, jammed his spurs to the horse's sides, and with a booming "Good-bye!" raced away at full speed into the vast space of Nebraska Territory toward the distant Missouri.

Hardly a minute had passed from the arrival of the Pony Express messenger until the hoofbeats of his horse once more were raising clods of dirt along the path worn in the ancient prairie turf. The lone rider pressed ahead toward another station twenty-five miles to the east, where he would pass his burden to a waiting relief rider. They were links in the chain of horsemen who, with daring and endurance, formed a precarious fast mail route between the United States east of the Missouri River and the remote American outpost beyond the Sierra Nevadas on the far Pacific shore, California.

Fast mail it was, indeed, by the standards of the day; far faster than had ever been achieved before. From San Francisco to St. Joseph in ten days. From there the letters could reach New York by train in another four days, fourteen days altogether. Until the Pony Express was started, the fastest anyone could hope to get a message across the great spaces to the western end of the railroad was twenty-three days, by way of the Butterfield Overland Stage that rattled and shook along a wandering trail through the Southwest to the tiny Mexican-American settlement of Los Angeles.

In those days men lived without electricity and all the wonders

Butterfield's Overland Mail Coach starting out from Atchison, Kansas

it was to bring. Railroad tracks had not yet been laid west of the Missouri River; the first chugging automobiles still were nearly forty years in the future. And the idea of men flying in motor-driven planes? Ridiculous! The Wright brothers had not yet been born.

West of the Missouri, the land belonged to the Indians, tribes that had roamed there for many hundreds of years—Sioux, Cheyenne, Pawnee—while hardly leaving a mark on the earth, such nomads they were. It was a world empty of the accustomed conveniences for the white men who invaded it: no fences, no bridges, roads that were only trails made by the Indians and enlarged by the wagon wheels of the early emigrants. The plains and the mountains beyond them spread across tens of thousands of square miles of untouched land, across which blew blizzards in winter and windstorms in summer to hamper the traveler. On the maps of the white man the land west of the Missouri was marked "Great American Desert." That was how little the newcomers knew about the country into which they were beginning to push.

Far to the west lay California, newly made a state of the Union but so far removed from the homeland back East that it was almost another world. Since the discovery of gold in 1848, thousands of adventurers had gone there by sea in sailing ships, either by the long dangerous voyage around Cape Horn or across the fever-infested jungle of the Isthmus of Panama, where they changed ships. Others, preferring to go by land, plodded across the Plains in covered wagons. No matter which route they chose, the emigrants cut themselves off from contact with the world back East. Time lost its meaning, one day turning into another as they drew farther away from the homes they had known.

The enormous rolling plains, the mountains, the desert, and then the final barrier of the towering Sierras before they reached California could be challenged in only two ways—by a man's two feet, or by the four-footed animals he rode or drove.

Jim Moore was among the dashing crew of young frontiersmen, some of them only seventeen years old, who carried the mail for the "Pony"—men like "Buffalo Bill" Cody and "Pony Bob" Haslam. Their riding and endurance were remarkable. Through the night they rode, alone in the blackness on a skimpy trail, no other human within miles. Through thunderstorms, through snowdrifts, sometimes fighting off Indians, they carried their cargo of the written word between California and the rest of the United States. Only one city worthy of the name existed in the run of almost two thousand miles. That was Salt Lake City in the Great Salt Lake Valley of Utah, where Brigham Young presided over his colony of Mormons.

It was Jim who set one of the most famous Pony Express endurance records. From Midway Station in the Nebraska prairie to the cluster of shacks known as Julesburg, Colorado, was 140 miles. Carrying the mail westbound, Jim reached Julesburg to find the rider for the eastbound trip so sick that he could not ride. Hardly had he dismounted than the Pony messenger from the west was sighted. Somebody had to replace the sick man in a hurry.

Wiping the sweat from his face, Jim looked around at the cluster of men outside the Pony Express office. No one else was young enough or strong enough for the job.

"I'll do it," he volunteered.

He gulped a mug of cold coffee, stuffed cold meat into a pocket

Pony Express rider gallops across the Plains to escape pursuing Indians, past an elevated Indian burial site.

for a meal en route, and headed back east over the trail he had just finished. His slender figure was bent forward in the saddle near exhaustion when, many hours later, he rode into Midway Station and passed the *mochila* to the next man. He had ridden 280 miles in twenty-two hours, almost around the clock.

Today, several months later as Jim galloped along on his regular run, the route was not so lonely. In these summer weeks the emigrant traffic to the west was at its height, the parties moving as rapidly as they could in order to cross the Sierras before the winter snows.

Coming toward him he saw a horseman and then, off to the left, another. Were they Indians—scouts for a hunting party like those he had evaded earlier? His sharp frontiersman's eyes soon convinced him that they were not. These riders sat in their saddles like white men. In the distance behind them he saw the canvas tops of a wagon train, nearly a hundred vehicles approaching him in four lines abreast, looking not unlike an armada of small sailing ships on the open sea. The horsemen were hunters for the emigrant party, seeking a herd of buffalo and at the same time keeping a wary eye for Indians.

Soon Jim drew abreast of the wagons. There was no stopping for a chat; the precious minutes slipped away too quickly. He smiled at the signs painted on the canvas of some wagons: PIKE'S PEAK OR BUST. With a sweeping wave of his hat and a shouted "Hello!" that was half lost in the prairie wind, the rider clattered past. He was alone again.

"What slow traveling that must be," Jim thought. Three months for a wagon train to cover the ground that the flying hoofs of the Pony Express did in ten days! The covered wagons creaked so slowly behind their teams of mules or oxen that the travelers could walk as fast as the wagons moved. He remembered the emigrants he had talked to in Julesburg recently who said their overloaded train was making only ten miles a day.

How fantastic it would have seemed to Jim if someone had told him that a hundred years in the future people would consider his own reckless gallop hardly more than crawling, that broad twin strips of concrete would unwind like white ribbons along this very route. The highway signs would say Interstate 80, and an automobile would travel as far in one hour as the wagon train he had just passed could go in four days.

Glancing back over his shoulder, he saw the signs of a gathering thunderstorm. It would be a race with the storm to the next Pony Express station. These summer storms plagued the Pony riders, who had to keep galloping no matter what the weather. Even if a rider stopped on the open prairie, no shelter could be found most of the time. The coming of a thunderstorm could be awesome. The riders had learned to watch for signs on a hot afternoon: billowing white clouds banking up like a lofty column of cotton in the blue western sky. Soon these were altered into a blue-black mass which spread across the entire sky, punctuated with lightning streaks jabbing their fatal white fingers down toward the earth. Thunder which began as a distant rumble soon came in bolts so booming that the vibration rattled the canvas of the wagon trains.

Docile streams turned into torrents. The Pony Express riders had

Sudden thunderstorm on the Plains overturns covered wagons and sends horses running frantically in search of cover.

to ford these swollen rivers, usually without help because they rode alone. Swimming a horse across a high stream in the dark of night without a light or another human for companionship was a test of a man's determination. Leather pants and jackets became soaked during the crossing, but the oil silk wrappings kept the letters safe inside the *cantinas.*

Even so, the threat of a gathering storm added zest to the long dreary hours a Pony Express rider must travel. So did the herds of buffalo that roamed the Plains. Today, as on most of his trips, Jim saw small groups of the shaggy, humped animals grazing on the horizon. For the hunters attached to the wagon trains, the sight meant the prospect of fresh meat, but not for the Express riders. The guns they carried on their saddles were for self-protection in case of Indian attack, or if by chance a rider was caught in the midst of a buffalo stampede. That could be a terrifying experience.

One traveler whose covered wagon had crossed Nebraska not long before wrote to friends back East: "I have seen the plain black with buffalo for several days' journey as far as the eye could reach." The herds, seeking water in the rivers, came north along the broad dusty paths that crossed the Pony Express trail. They plunged into

Old print shows herd of bison, erroneously called buffalo. The Plains were once the home of great herds of bison.

Today a bison herd roams in the Theodore Roosevelt National Memorial Park, thanks to conservationist efforts.

the streams by the thousands and swam across, fouling the water.

Jim remembered sitting outside the Pony station at Julesburg, listening to the drivers of a wagon train telling how a herd of buffalo had borne down on their caravan like a tornado, almost at this very spot. The front of the galloping mass was half a mile wide.

"Quick, get out of their path!" The word raced back from the wagonmaster of the threatened train. "Run for it!"

Whips cracked. Reins snapped. Drivers shouted. The wagon train gathered speed until it came as close to a gallop as the oxen and mules could go. The earth shook from the hoofbeats of the wild animals thundering shoulder to shoulder toward the train.

For several minutes the herd looked as though it would crash into the center of the train, smashing the wagons to splinters. But the disorganized caravan's scramble was just fast enough. The edge of the buffalo mass galloped headlong behind its rear, missing the last wagons by only a few yards.

During the time of the Pony Express, it is estimated that more than twenty million buffalo roamed the Plains. To Jim Moore and his fellow Express messengers, buffalo were as common a sight as are cows behind the fences along a modern highway. But with their senseless greed to kill far beyond their needs, the white men within a few decades almost wiped the buffalo out of existence.

The rumble of thunder came louder to Jim's ears. A jagged bolt of lightning zigzagged across the sky to the southwest. He calculated the distance to the next Pony Express station, looked over his shoulder at the gathering storm, and muttered to himself, "Good, I'll make it." The Indians had given him enough surprise and trouble for one day; he didn't want more.

Something startling awaited him beyond the slight rise in the ground ahead, however; something that would change his life and that of others accustomed to the vast silence of the Plains. As he topped the crest, the Pony Express messenger blinked in surprise. Reaching westward toward him like a slender finger was a line of holes dug in the matted turf. Beyond them rose rough-hewn poles in single file, each perhaps a dozen feet high with a crossbar near its top.

Here was something astonishing: wooden poles stretching across

Pony Express rider salutes the telegraph line, while an emigrant wagon train in the background plods slowly along.

a land where there were no trees. Men were planting poles in the holes, heaving them from a stack on a wagon. Soon the rider reached a gang of workmen who were stringing a strand of wire to the crossbars. The single wire was spun out as far to the east as Jim could see, and extending all the way to the Missouri River.

Jim was too curious to resist the temptation to stop.

"This the telegraph I've heard about?" he asked.

A workman laughed. "That it is. Better go put that pony in the barn. You'll soon be out of a job."

A new marvel had come to the Plains. That single wire charged with electricity was about to tie the world of the West together.

CHAPTER 2

The Telegraph Is Born

The idea that led to the electric telegraph and brought men at distant points into instant contact was born not in a scientific laboratory but on the deck of a sailing ship named the *Scully* in the middle of the Atlantic Ocean. The man to whom it occurred hardly seemed the type to be an inventor. He was a portrait painter named Samuel Finley Breese Morse, and in that year of 1832 he was among the most famous artists in the United States.

Morse had been in Paris painting and was homeward bound to New York. Aboard the slow sailing ships the hours were long, and the passengers amused themselves by talking and theorizing about many topics. At the time, man's knowledge of electricity was skimpy in the extreme. This was nearly a half century before the electric light was invented. Some men had begun to realize, however, that, if harnessed, electricity had tremendous possibilities.

Mixed with his artistic sensitivity, Morse had a practical turn of mind. Talk about electricity intrigued him. One day, the shipboard conversation turned to the recently invented electromagnet. This was a piece of soft iron bent like a horseshoe, around which wire was wound. When electricity was sent through the wire, the iron bar became a magnet and could pick up metal objects.

A companion asked if the flow of electricity did not diminish when the wire was long. No, said another, electricity passed instantly over any known length of wire.

Morse remarked, "If this be so, and the presence of electricity can be made visible in any desired part of the circuit, I see no reason why intelligence might not be instantaneously transmitted by electricity to any distance."

Years were to pass before Morse developed the telegraph to its commercial stage, but the basic idea was born in his mind at that moment. Indeed, although he didn't know it until later, the idea had occurred to others earlier, but none of them developed it into the practical form that Morse did, with the help of several partners and assistants.

The basic theory is this. A wire is strung between two points, perhaps many miles apart. A key at one end is pressed down by the hand of the operator, closing the electrical circuit and sending a pulse of electricity through the wire. When the key is released, the circuit is broken. If the key is pressed down and released almost immediately, the brief pulse of electricity sent along the wire is a dot. If the key is held down three times as long, the pulse is a dash.

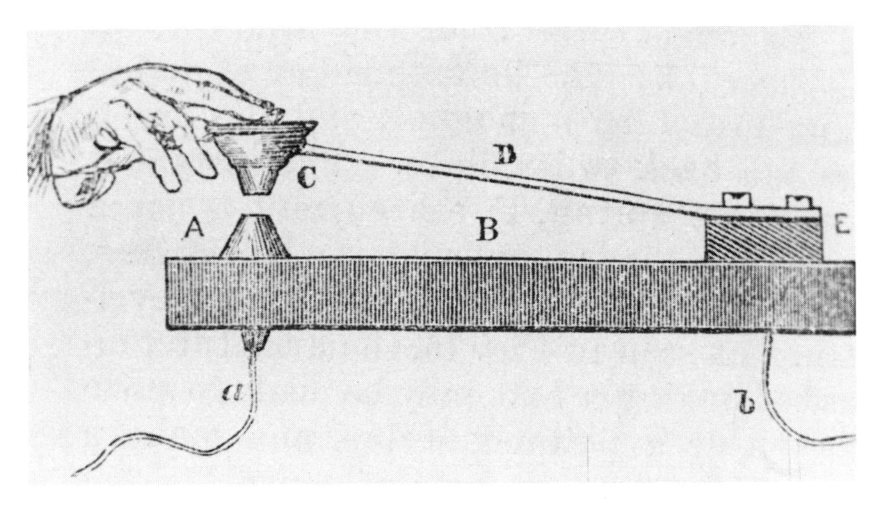

This drawing explains the principle of the Morse telegraph key. Electric current enters the little anvil, A, through wire a *that passes through the wooden table, B. Pressure of the button, C, at the end of the spring, D, against the anvil closes the circuit. The current flows into the metallic block, E, and into the departing wire,* b.

When arranged in certain combinations known to the operators at both ends of the wire, called the Morse Code, these dots and dashes spell out words. Morse Code looks like this:

A	.-	N	-.
B	-...	O	---
C	-.-.	P	.--.
D	-..	Q	--.-
E	.	R	.-.
F	..-.	S	...
G	--.	T	-
H	....	U	..-
I	..	V	...-
J	.---	W	.--
K	-.-	X	-..-
L	.-..	Y	-.--
M	--	Z	--..

At the receiving point, Morse made use of the electromagnet that he and his friends had discussed on the *Scully* to build a sounder. The incoming pulse of electricity surged through the wire wrapped around the horseshoe-shaped iron bar. This magnetized piece drew up to itself one end of a bar on a lever, causing a click that formed a dot or dash depending upon how long the pulse of electricity lasted. When the sender released the key, stopping the electricity, the magnetic action ended and the recording bar returned to its original position. At intervals along the wire, a special receiving instrument moved the signals through a relay battery, which renewed their strength.

Although the principle seems simple today, in the late 1830's many men refused to believe that a strand of wire thus could be made to "talk."

Fascinated though he was by his crude idea for a telegraph, Morse had to earn money to live on, so he kept on with his painting. He also ran unsuccessfully for mayor of New York City. In 1837 he developed his telegraph idea far enough to test it. He strung seventeen hundred feet of wire around his room at New York University, where he taught. It worked; his signals were transmitted from one end of the wire to the other. Early in 1838 he went to

Washington and demonstrated the telegraph to members of a Congressional committee in a room in the Capitol. Stringing ten miles of wire around the room, he made the signals perform just as he had expected. President Martin Van Buren and his Cabinet came to inspect the invention.

Surely the country's most important men, having seen the telegraph, would recognize its tremendous potential and vote Morse a federal subsidy to build a line between cities, or so he thought. But he had only disappointment. The members of Congress decided that Morse's invention was just a "thunder and lightning 'jim crack,'" as one man called it sarcastically. They wouldn't put up a penny.

Six years passed before Morse had another opportunity to demonstrate his telegraph, by now somewhat improved, to Congress. This time he asked for $30,000 with which to build a telegraph line from Washington to Baltimore, Maryland, forty miles away. Stringing his wires from one room of the Capitol to another, he demonstrated to the dubious Congressmen how his "talking wire" worked.

So doubtful were some Congressmen that they laughingly tried

The House of Representatives in session

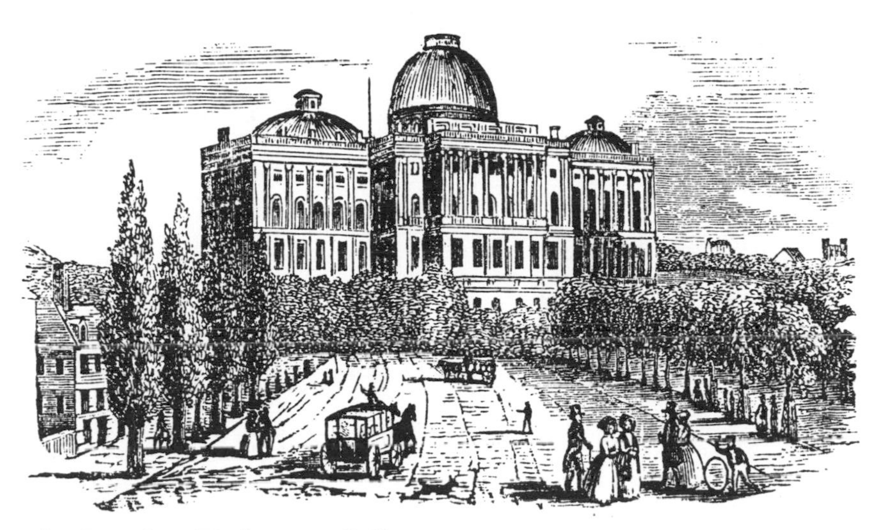

The Capitol at Washington, D.C.

to load the bill authorizing the $30,000 with nonsensical amendments.

"If we are going to spend money to develop the telegraph, I move that we also appropriate money to investigate mesmerism," one House member said, referring to what is now usually called hypnotism.

"We also should put up money to test Millerism," another joker asserted. Millerism was a religious sect which predicted the second coming of Christ in 1844, the following year.

A vote actually was taken in the House to add mesmerism and Millerism to the telegraph bill. Fortunately, better sense prevailed and the amendment was defeated. Eventually the House passed the bill. So did the Senate in the final hours of its session, after Morse fell into a state of nervous despair because he feared the measure was going to be lost in the rush toward adjournment, as it almost was. At last the United States was to see the telegraph in operation!

It would, that is, if the system could be made to work. Stringing miles of wire around the rooms of a building was one thing; making the electrical wire transmit messages for long distances across country was quite another matter. Indeed, Morse's wire to Baltimore almost failed. He tried to lay the wire underground along the right-of-way of the Baltimore and Ohio Railroad, inside a pipe. After several miles of wire had been laid, Morse discovered to his dismay that the telegraph line would not work because the insulation of the wire was defective and so the electrical circuit failed.

With only $7,000 of his $30,000 left, Morse pulled up the underground wire and installed it on poles instead. The poles that his men dug into place along the railroad track were the first of thousands upon thousands that soon would be stretching across the built-up parts of the United States and into the wilderness of the West.

Soon Morse had a dramatic opportunity to show the doubters in Washington that his telegraph really worked. In that spring of 1844 both the Whig and Democratic Parties held their national conventions in Baltimore to nominate candidates for President and Vice-President of the United States. When the Whigs met on May 1, Morse's telegraph line had reached Annapolis Junction, about twenty-two miles from the national capital.

Morse sent his assistant, Alfred Vail, to Annapolis Junction. When the train from Baltimore to Washington carrying delegates home from the convention reached the Junction, Vail asked who had been nominated. "Clay and Frelinghuysen!" a delegate shouted.

The selection of Henry Clay for President had been expected, but the choice of the little-known Theodore Frelinghuysen for Vice-President was a surprise. Vail clicked out the news on the telegraph and Morse received it on the instrument he had set up in the Capitol.

Reaching Washington, the delegates hurried to the Capitol to tell their news. They were astonished to hear their listeners say, "We know that already." Morse's telegraph had beaten them.

In another three weeks, the telegraph line had been completed to Baltimore. One end was in the Supreme Court chamber in the Capitol—this was before the Court had separate quarters of its own—and the other was in the railroad station at Baltimore. On May 24, 1844, Morse invited a group of high level political figures into the small Court chamber to witness the sending and receiving of the world's first formal telegraph message. Among those crowded

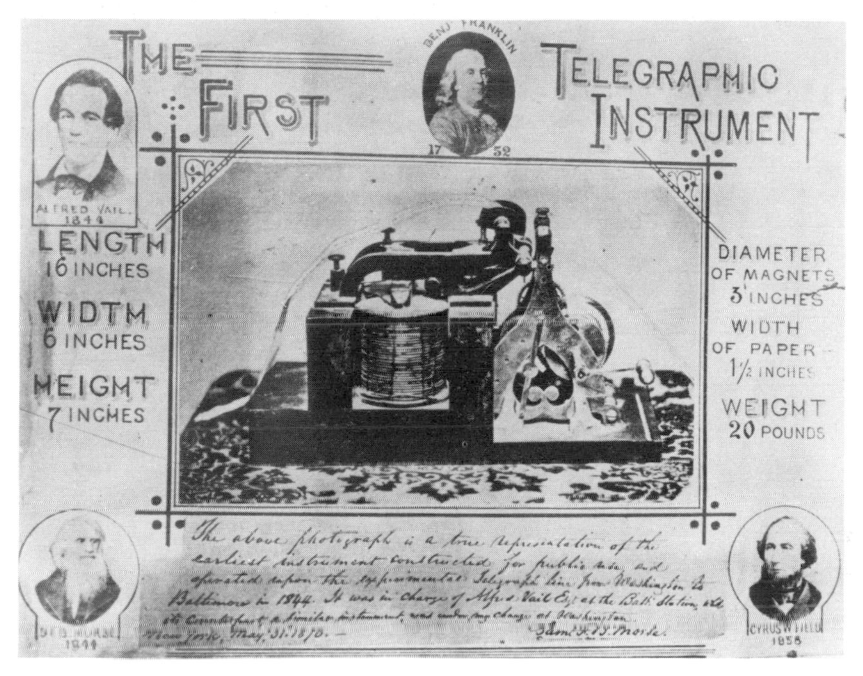

Photograph of the earliest Morse telegraph instrument used on the Washington-Baltimore line in 1844 by Samuel F. B. Morse. Descriptive material concerning it is in Morse's handwriting.

around the instrument at which he sat were Henry Clay, the Presidential nominee, and the famous Dolley Madison, wife of former President James Madison.

The honor of sending the first telegram was given by Morse to Annie Ellsworth, daughter of a government official who was a longtime friend of the inventor. So there could be no possibility of cheating, Morse had no advance knowledge of what her message was. It proved to be words from the ancient soothsayer Balaam which Miss Ellsworth and her mother thought especially suitable for the new electrical magic.

In code, Morse tapped them out: "What hath God wrought!"

Vail in Baltimore received the words, then transmitted them back to Morse. The witnesses in the Supreme Court chamber cheered as they came clicking in.

After years of poverty and disappointment, Morse suddenly was

W h a t h a t h G O D
w r o u g h t.

The first telegraphic message sent by the Morse system—"What Hath God Wrought."

a hero. His triumph became even greater a short time later when the Democrats held their nominating convention in Baltimore. Reports on the proceedings were telegraphed to Washington and were heard eagerly, much as people today gather around television sets to see what the presidential conventions are doing. After nominating James K. Polk for President, the convention chose Senator Silas Wright of New York for Vice-President. News of his nomination was telegraphed to Wright in Washington.

A few minutes later Wright handed a message to Morse and said, "Please send this to the convention."

His telegram stated, "I refuse to accept the nomination." The convention was so startled that it dispatched an official delegation to Washington by train to confirm that Wright actually had sent the message.

Now the telegraph was a success, for certain. Morse offered to let the federal government buy the rights to the telegraph for $100,000 and run it as a government business. His offer was turned down. Officials didn't think the business could make a profit, one of the classic miscalculations in American history. Quickly, private investors licensed by Morse and his associates erected wires from city to city in the eastern United States and set up a myriad of telegraph companies. The Western Union Telegraph Company, which bought up the small companies one by one and developed a far-reaching combined system, soon became a money-maker of gigantic proportions.

Section by section, the wires were extended between cities, up and down the Atlantic coast, moving both inland and through the South to New Orleans. Usually the wires were of copper, the best conductor of electricity. Galvanized iron was used, too, because it was cheaper, but it has higher resistance to the electric current and tends to corrode, except in a dry climate. On the wooden crossbars of each pole, the wire was fastened to an insulator, usually made of glass. These insulators were mounted on wood or metal pins set vertically into the arm.

Excitement was great as the telegraph wires reached each community. Those towns and cities that were not yet on a railroad line—and there were many of them in the 1840's and 1850's—suddenly found that they could communicate with the rest of the country at a speed faster than a horse's pace. Promoters made towns pay them sums of several thousand dollars before they would run the wires through those towns.

So many short-distance companies existed that a long-distance message, say from New York to New Orleans, went part way on one company's wires, then had to be carried to the office of another company for forwarding another part of its journey. At first messages were received with dots and dashes on a long moving strip of paper, recorded by a pencil attached to the receiving arm of the instrument, then transcribed into English letters by the operators.

Since this method was laborious, operators took to copying the English letters directly onto message blanks as they heard the groups of dots and dashes coming in. Soon the paper tape method was abandoned. Reliance was placed entirely upon the ability of the operator's ear to distinguish the dots and dashes, and upon his brain to translate the staccato flow of sound into letters of the Morse Code. When the typewriter came into use later, the flow of dots and dashes into the operator's ears emerged from his fingertips as the appropriate letters typed on a message blank.

Among the first operators in the country to receive messages directly by ear was a 15-year-old boy in Louisville, Kentucky, named Jimmy. Crowds came to the telegraph office and stood in awe while

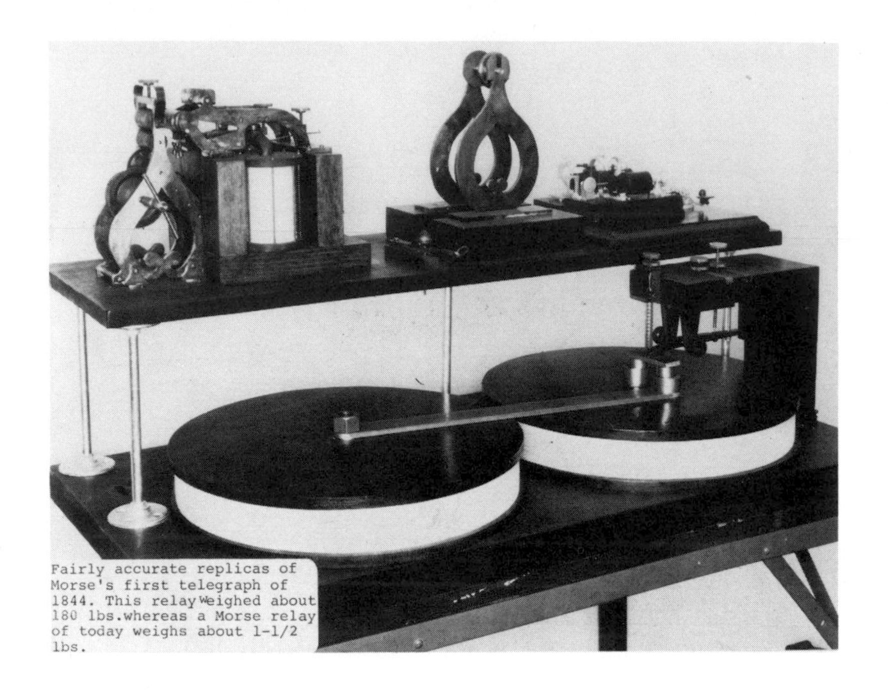

Fairly accurate replicas of Morse's first telegraph of 1844. This relay weighed about 180 lbs.whereas a Morse relay of today weighs about 1-1/2 lbs.

Replicas of Morse's first telegraph equipment

Jimmy wrote out incoming messages from the flow of sounds. Jimmy's fame spread to the point that the flamboyant showman P. T. Barnum offered him a job in Barnum's American Museum in New York. Jimmy refused, not wanting to leave Louisville. He was killed while a soldier in the Civil War fifteen years later and his gravestone was decorated with telegraph lines.

Not all operators were as skilled as Jimmy. So many errors occurred that stories about the blunders became the same kind of jokes as today's tales of slow and misplaced mail. At times the operators were too careless to check back by wire to the sending operator for confirmation.

A story in *The New York Times* of September 29, 1860, is typical: "A gentleman who prefers to be anonymous, for obvious reasons, returning home from Norwalk, Connecticut, a few days ago by railroad, sent a telegram in advance, ordering a horse and carriage to meet him on his arrival. What was his surprise on reaching the depot to find a hearse and carriage in waiting. This illustrates the intelligence of many of the telegraphic employees."

In those days when attendance at public school was often casual, boys no older than Jimmy held full-time jobs as operators. Despite the slurs at the intelligence of operators, many men who later became famous worked as telegraphers in their youth, including

Thomas A. Edison and Andrew Carnegie. At the time of the Civil War, a telegraph operator worked ten or fifteen hours a day, six days a week, and often was called on duty on Sunday as well. For this he was paid $70 or $80 a month. Nobody even thought about overtime pay. When an operator had to sit that long at a stretch, concentrating on the dots and dashes, it isn't surprising that he made mistakes.

Transmission of messages by Morse Code averaged about twenty-five or thirty words a minute, depending on the skill of the sender and receiver and the clarity of the circuit. Often the wires broke down. Pioneer operators also had the responsibility of maintaining the batteries in their offices that provided the electricity for the circuits. There was no generation of electrical current in the country in those days. Young Tom Edison, the telegraph

Old print of Chicago shows telegraph lines strung through the streets.

operator of 1863, didn't invent the electric light until 1879. So the telegraph companies had to create their own electricity with cumbersome batteries using sulphuric acid.

Slowly the telegraph lines were extended westward, coming to the raw frontier towns east of the Mississippi River including Chicago, then to St. Louis on the western bank of the Mississippi, and eventually across Missouri to the east bank of the Missouri River. When the Civil War broke out in the spring of 1861, the land beyond the Missouri River was the great mysterious West. The river divided the populated part of the United States, sparse though that population was in many areas, from the beckoning wilderness of the Great Plains and the towering mountains.

A telegraph wire was strung across the Missouri River to Brownsville, a settlement in Nebraska Territory on the west bank, near the end of August, 1860. What jubilation its arrival created! A salute of thirty-four guns was fired by the inhabitants in honor of the first telegraph to reach the Territory. Bonfires lit the sky that night and fireworks exploded. Enough instruments were rounded up for an impromptu band to blare out march music, after which, inevitably, the local politicians made speeches. The thread of copper wire joining Brownsville to the States created a feeling of wonder and mystery. Men from the settlement, and others who had come on horseback from sod huts on the prairie to witness the machine, stood silently with hats held across their chests in a reverent attitude as though they were praying, while the operator at its table with its chattering key copied messages dispatched from New York less than an hour earlier.

There was a sense of marvel even in the far more sophisticated offices of the New York newspapers. *The New York Times* published an editorial in praise of the achievement, saying, "A new line of wire stretches away to the confines of civilization, bringing the seaboard in instant contact with the Far West, and tying the newest and tenderest shoot of our family tree to the parent-stem. Considering the novelty of the occasion, Nebraskans must be pardoned if they are a little vain of their new advantage. They have actually come into possession of a magnetic battery of their own, with all its paraphernalia of pots and acids; they possess certain miles of

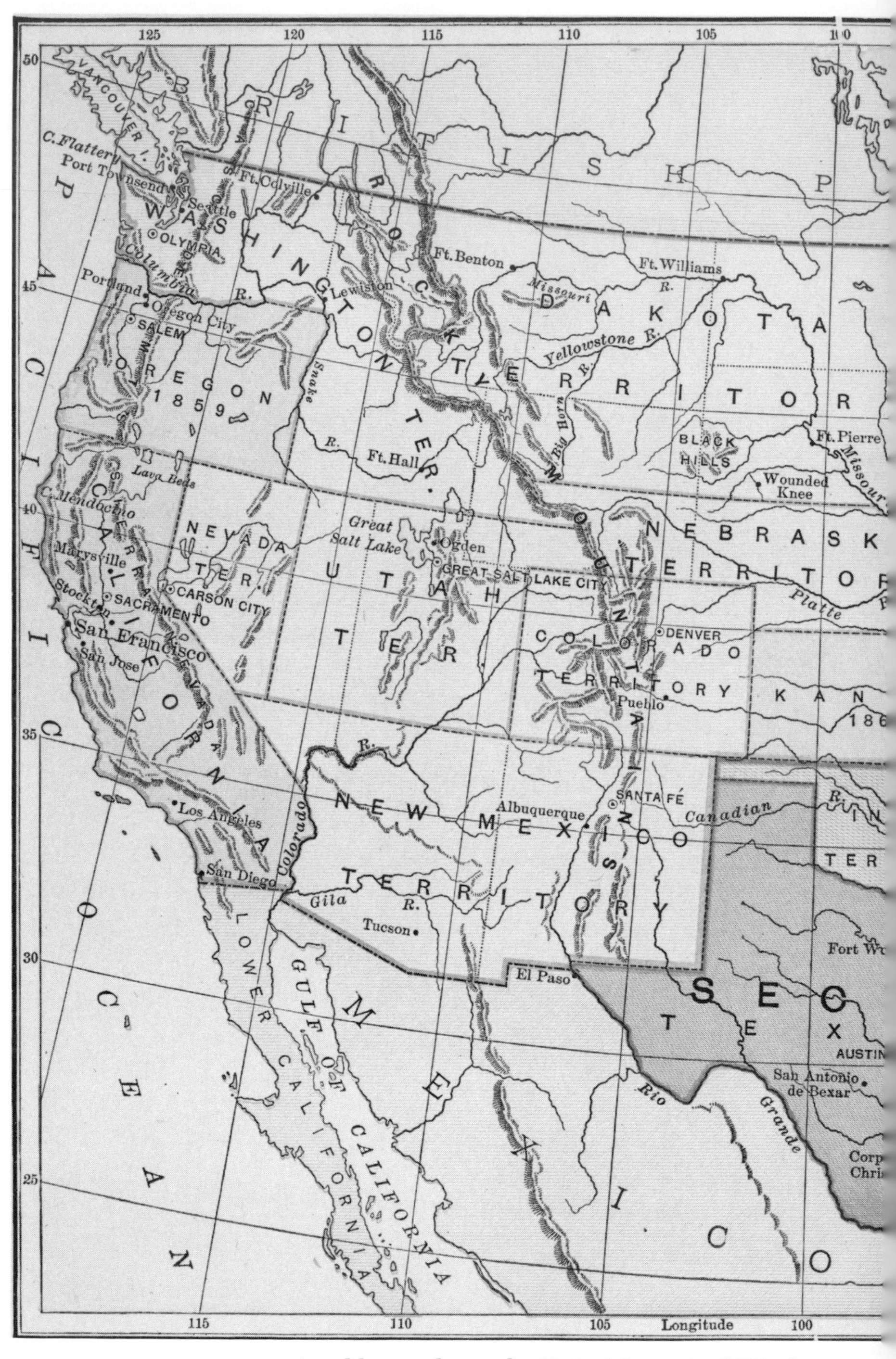

An old map shows the United States in 1861, the

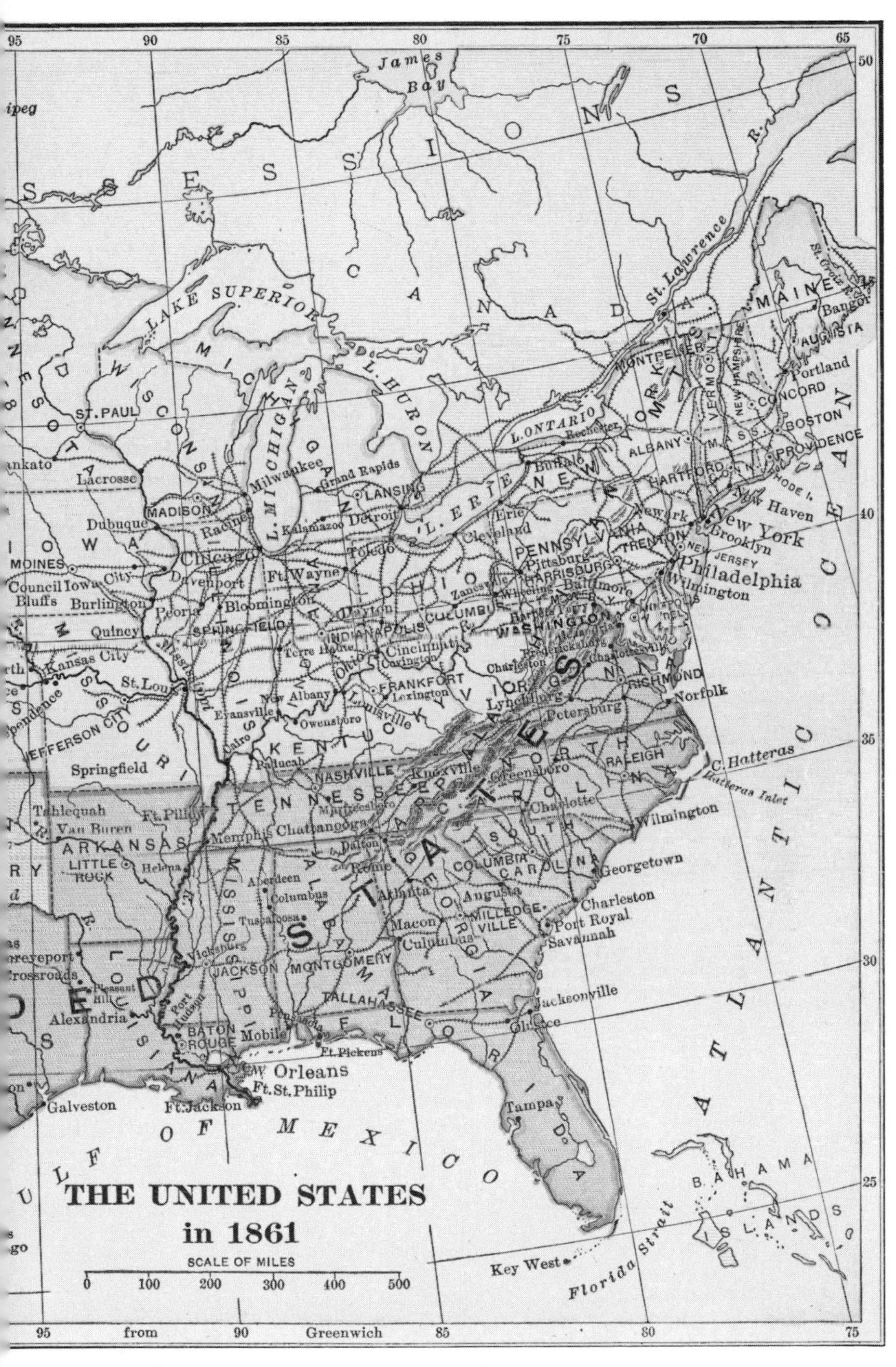

year after the telegraph crossed the Missouri River.

copper wire and a telegraphic office and probably an operator, who is not more stupid than the average of such functionaries . . . and who will commit no more blunders than he can possibly help, so as not to vex the righteous soul of the Nebraska community beyond human endurance." (How that *Times* editor hated telegraph operators!)

"With the extreme East and the almost extreme Far West thus linked fast together, what more do we need to cement the tie of brotherly affection?" After reciting how much had been accomplished in the sixteen years since Morse successfully sent his first telegram, the *Times* concluded, "In September, 1860, we have a continuous line from Sackville, Nova Scotia, to New Orleans, and another westward to the half-peopled wilds of Nebraska."

Had the editor but realized it, the most exciting part of the telegraph story still was to be told. From his desk in New York, the Missouri River at Brownsville appeared to be the Far West; yet nearly two thousand more miles of telegraph poles and wires remained to be installed across the wildest, broadest part of the continent before the telegraph reached the western shoreline in California. Brownsville wasn't the Far West, but merely the front door to that challenging wilderness.

CHAPTER 3

Gold, Ponies, and Wires

For a long time the people of California and those of the early-day United States had nothing to say to each other. In fact, each country was barely aware that the other existed, even though they were on the same continent. So it didn't matter how many months were required to send a message across the huge empty space between them.

California was owned by Mexico. The easy going residents of its ranchos and pueblos cared little about the world east of the lofty Sierra Nevadas. Their minds and hearts turned south to Mexico City, not toward the young country three thousand miles to the east.

When an occasional American sailing vessel came around Cape Horn and up the Pacific Coast, the Californians looked upon the English-speaking sailors uncertainly, not quite sure what to make of their exuberant ways. Americans building their country on the Atlantic Seaboard had much the same attitude about California. To them, "out West" meant the infant frontier towns of Chicago and St. Louis. Between the States and California lay a great void.

Almost until the middle of the 1800's, the Mississippi River was the western boundary for most life in the United States. A few pioneers pushed across Iowa and Missouri but stopped when they reached the Missouri River. Only a handful of mountain men, fur trappers and explorers searching for pelts, ventured west into the Rocky Mountains. These men were individualists who expected to be out of touch with their families for years and in many cases

An old print from an 1860 schoolbook on geography shows the artist's conception of life in California during the Gold Rush days.

probably cared little. Indeed, some were pleased that certain people back in the States didn't know where they were.

Then in the 1840's things began to happen.

In 1845, a year after Morse sent his telegram from Washington to Baltimore, fewer than seven hundred Americans lived in California. The Mexican rulers of California were weak and lazy, however, and these Americans were eager. They staged what became known as the Bear Flag Revolt in 1846 and declared California's independence. Two years later, Mexico formally gave California to the United States.

Almost simultaneously, men discovered that California had something the whole world prized—gold! James Marshall made the discovery at Sutter's Mill in the central California mountains in January, 1848. Communications were so slow, however, that word of the discovery didn't reach New York and Washington for seven months.

Gold fever swept through the United States like an epidemic. When spring came in 1849, men set out by the thousands on the

Old print of miners washing out gold

Below: *San Francisco, an old print. The circled shack was the only human habitation in the Bay area in 1835.*

long trek to make their fortunes in the mines, a dream that came true for only a few of them. They hurried west by sea and by wagon train. During 1849, about eighty thousand men and a few women reached California, three-fourths of them Americans. The tall masts of five hundred sailing ships rose in San Francisco Harbor.

Many of these ships were abandoned by their crews, the sailors heading for the gold diggings as fast as their feet or a mule could carry them.

The Forty-Niners had their hearts and families back in the States. Not being loners like the mountain men, they were anxious for news from home. They didn't want to wait three or four months for a letter or to send back word about their own adventures.

"Do something about getting us better mail service!" they demanded. But their request took a long time to reach Washington and was little heeded when it arrived.

Although separated from the main body of states by more than fifteen hundred miles, California was admitted to the Union as a state in September, 1850. With senators and congressmen representing her in Washington, she had a better opportunity to press her case. These delegates were on their own with little advice from the folks back home, since a message to California and a reply to it required four or five months.

The excitement of the headlong western migration of the 1850's was to be overshadowed a decade later by the even greater drama of the Civil War. The storm was beginning to gather when California entered the Union. She was admitted as a free state, one in which the holding of slaves was illegal. While this seems like the most natural thing in the world today, in 1850 it was the subject of bitter dispute. The evil practice of slavery flourished in the South, and southern politicians carried great weight in Washington. This fact was to influence heavily California's efforts to get faster mail, a telegraph line and eventually a railroad.

The southern politicians wanted any government-subsidized transportation route to the Pacific Coast to go through the southern and southwestern states, rather than directly across the "free" central part of the country. An effort by a private operator to run a mule-drawn coach service with mail east from central California via Salt Lake City in 1851 encountered such poor traveling conditions that the first trip from Sacramento to Salt Lake City required fifty-three days. A man could almost walk the distance in that time.

In the previous year—September of 1850—Congress had created the Utah Territory. News of this event, so important to the Mormon settlement in Salt Lake City, didn't reach there until the following January—four months en route. It was carried by way of Panama to San Francisco, then eastward again across the mountains and desert to Salt Lake.

Finally, in 1857, the federal government agreed to help pay for a stagecoach service from the States to California, ending at San Francisco. This was the Butterfield Overland Mail, an enormous project along one of the craziest routes anybody could devise. The Southerners in Congress got their way in designing the route. The decision was made by Secretary of War Jefferson Davis, who later became President of the Confederate States of America. Little wonder he chose the route he did.

This Overland Mail route ran 2,700 miles in a great southern-bending arc from St. Louis, Missouri, on the Mississippi River south to Little Rock, Arkansas; west to El Paso, Texas; on to Yuma, Arizona and Los Angeles, California (still a tiny community enormously overshadowed by San Francisco); and four hundred miles north up the California coast to San Francisco. A trip aboard a swaying, bounding Concord stagecoach over this route took twenty-three days, traveling night and day through unpopulated areas. Each Overland Mail coach, drawn by a six-horse team, carried

Passengers on Butterfield Overland Mail Coach prepare a meal during a rest stop on their jolting trip across the Plains.

six passengers and four sacks of mail. To those passengers whose stomachs and backs were strong enough for the trip, the twenty-three days over the rutted trail seemed like a lifetime. Receiving mail from St. Louis in twenty-three days caused jubilation in California, however. Imagine, a letter from home in just over three weeks!

Soon, however, this improved speed wasn't good enough. Demand grew for faster, more direct mail service.

That is when the Pony Express was born. Its operation, daring and fascinating, became one of the most romantic stories of the Old West. Yet the truth is that the Pony Express ran for only nineteen months, from April, 1860, until October, 1861. Its early death was foreshadowed even before the first rider galloped onto the trail by the lengthening fingers of the telegraph lines.

A rickety telegraph line whose single wire often was fastened to trees or shrubs had been strung in 1859 across the Sierra Nevadas along the emigrant trail from Sacramento to Placerville (which had abandoned its earlier name of Hangtown), past Lake Tahoe high in the mountains and down the eastern slope of the Sierras to the community of Carson City, which later became the capital when Nevada was admitted to the Union as a state. Frederick A. Bee built the line, which became known as "Bee's Grapevine," a reference to the old joke about news traveling by the grapevine—that is, by word of mouth. The line was as wobbly as a grapevine, too. When it broke, which was often, travelers helped themselves to sections of the loose wire to repair the wheels of their wagons and other equipment. With straight faces, they pretended to think that somebody had left the wire there in the mountains for that purpose.

This line to Carson City was in operation before the Pony Express began. At the other end of the long Pony route, a telegraph line extended east from St. Joseph, Missouri, to New York. That left

A wagon train and a stagecoach travel winding route over the high Sierras near Carson City, Nevada, as telegraph poles cling to the side of the road. "Bee's Grapevine" first carried messages between the Nevada gold mines and California.

a gap of about seventeen hundred miles between telegraph outposts to be covered by the Pony Express.

Even though messages carried west by the Express riders could be transferred to the telegraph at Carson City, the riders traveled the remaining distance to San Francisco. Only the biggest news was put on the telegraph at Carson City. Transmission was slow, uncertain, and expensive over the line, and those who sent the letters at $5.00 a half ounce, an extremely high price in those days, wanted them delivered unopened at their destination.

The Pony Express was operated as a private enterprise without government subsidy by the freighting firm of Russell, Majors & Waddell. Brief as its life was, the Pony Express had many thrilling moments. Messengers rode like the wind. Traveling west in a stagecoach, Mark Twain saw a Pony Express rider gallop by so fast that the coach passengers hardly realized he had passed them "but for a flake of white foam left quivering and perishing on a mail sack."

A string of 190 stations, each usually only a single building of adobe mud, was established. Riding night and day, the riders in relay carried the cargo of feather-light letters about 250 miles in twenty-four hours when weather conditions were good. At peak speeds they made twenty-five miles an hour in bursts. In the heavy snows that buried the plains and the mountain passes, the riders were slowed almost to a walk. A relay station was supposed to be sited every ten miles, and most riders had routes of about fifty miles. Because of long waterless stretches in the desert and other dangers, the stations sometimes were farther apart than that, and riders like Jim Moore carried the mail often on rides of more than a hundred miles.

When word of great events back East was handed to them, the Pony Express riders made extraordinary efforts to break speed records. The ominous news that the Civil War had broken out with the Confederate shelling of Fort Sumter, South Carolina, reached San Francisco in eight days and fourteen hours from the moment the rider left St. Joseph, having been forwarded by telegraph from Carson City when the breathless rider reached there.

Hardly had the Pony Express started operations before war broke

The Pony Express station at Cheese Creek, Nebraska

out in Nevada between the Paiute Indians and the settlers in the huge arid land whose sagebrush plateaus and treeless alkali flats were broken only by a few winding rivers. The Pony Express route went through the heart of the fighting zone.

The land was harsh, the distances great. In midsummer the temperature frequently reached 110 degrees in the shade. The Paiutes led by Chief Winnemuca were a poor tribe that survived by gathering nuts, hunting and fishing a little, and digging roots. This led to the nickname "Digger" Indians. At first they were friendly toward the miners and ranchers who came to the Nevada region in the 1850's. They became angered by the way some white men treated them, however. When a Paiute brave discovered that two Paiute girls who had disappeared from the tribe's main camp at Pyramid Lake were held captive by white men at Williams Station, the Paiutes set out to rescue the girls and take revenge. The station was a trading post of huts on the California trail about sixty miles northeast of the new mining camp at Virginia City. Also, it was a relay station for the Pony Express.

Paiute horsemen swept along the Carson River bank into the station, shot five men living there and burned everything, riding

Old print shows a pack train loaded for a mining camp.

away with the horses and cattle. A Pony Express rider arrived at Williams Station, saw the result of the massacre, and raced on to Virginia City with the terrifying news.

His tale ignited an uproar in the mining camps and in nearby Carson City, the metropolis of the area with more than a thousand residents. Quickly a volunteer force was formed by Major William Ormsby to ride north across the desert to Pyramid Lake to punish the Paiutes. These avengers were untrained, poorly armed, filled with whiskey bravado, and convinced that they could defeat the Indians with ease. But the Paiutes under their war chief Numega outsmarted them. Ormsby's force rode into a cleverly set ambush. Showers of arrows and volleys of bullets fired by Indians hiding behind rocks and sagebrush cut down men and horses. Only a handful of the white men escaped, leaving seventy of their companions sprawled dead on the desert.

When the survivors struggled back to Virginia City, their story struck terror. All the little settlements on the Nevada side of the Sierras expected attacks by the Indians at any hour.

By then Bee's telegraph had been strung the additional miles eastward from Carson City to Virginia City, already showing signs of becoming a booming silver mining camp.

Operators in Virginia City and Carson City kept "Bee's Grapevine" chattering with dot-and-dash messages across the mountains to California, telling the news and pleading for help. Major D. E. Hungerford of the Sierra Guard of California happened to be in Virginia City at the time. He telegraphed to his headquarters in Downieville:

Virginia City, May 13, 1860

Send me immediately all the arms and ammunition of the National Guard. Telegraph Lieut. Hall at Forest City to send all the rifles in his possession. Send to Goodyear's Bar, to Captain Kinniff, to send me all his rifles. Forward as soon as possible. Big fight with Indians. The whites defeated. Send me your heavy saber. Spear, Meredith and Baldwin killed.

MAJOR D.E. HUNGERFORD

Tension was kept high by word that the Paiutes had been raiding Pony Express stations across Nevada. These isolated outposts were prime targets for the Indians. "Pony Bob" Haslam was one of the riders on the Nevada section of the route. Completing his regular run at Buckland's Station, a few miles from where the Indians had burned Williams Station, he found that his relief rider was too frightened of the Indian attacks to carry the mail on the next leg. The Pony Express superintendent was at the station. Recognizing the danger but anxious that the mail should get through, he said, "Bob, I'll give you fifty dollars if you will make the ride."

Haslam agreed. Mounting a fresh horse, he galloped east across the waterless white alkali flats and sandhills. At Cold Springs Station he changed horses, spoke briefly to the station keeper, and rode ahead to Smith's Creek thirty miles away, where he turned over the *mochila* to the next rider. Haslam rested for nine hours, then headed back carrying the westbound mail.

As he neared Cold Springs he let loose the traditional coyote yell to the station keeper, whom he had last seen fifteen hours earlier. No answering wave followed. Nor were any animals visible. Haslam reined in his horse at the building, puzzled by the silence. Then he saw the body of the keeper sprawled outside the station.

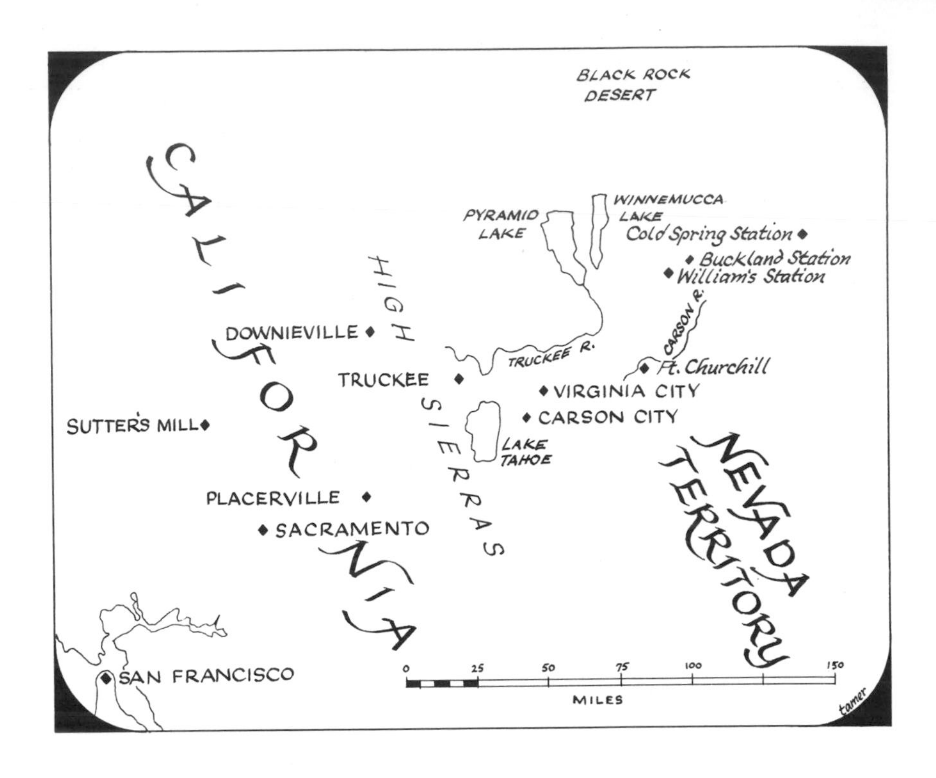

During Haslam's absence the Paiutes had raided the station, killed the keeper, and driven off the horses. Bob realized his own peril. The Indians probably were still nearby. He watered his tired horse, jumped back into the saddle, and raced to Buckland's with his tragic news; another Pony Express man was dead.

The risk was too great. Daredevils though they were, the mail couriers couldn't be exposed to further deaths. So the Pony Express superintendent shut down the service until a relief army being assembled back in the mining settlements could make the route safer. The thin line of communications between California and the States was broken; for ten days the Express didn't move in this portion of its long route.

From Virginia City and the other mining camps, a volunteer army of nearly six hundred men calling itself the Washoe Regiment was assembled, led by Colonel Jack Hays, a man experienced in the tricks of Indian fighting, and Major Hungerford. They were joined by 290 regular United States Army soldiers brought across the mountain trail from California. This time the military force that marched up to the Paiute headquarters at Pyramid Lake, unlike Ormsby's expedition, was well disciplined and aware of the Indians' prowess.

South of the lake, where the Truckee River's runoff waters from the snows of the High Sierras flow into the picturesque, rock-festooned body of water, the Indians waited. For a day, sharp fighting went on. Soldiers and Indians took shelter behind the rocks on the desert slopes and sought to mask their movements behind the skimpy cover of the gray-green sagebrush. By nightfall the soldiers clearly had the advantage. Because their leaders had been in fights with Indians before, they did not allow the men to be drawn into ambushes, and the Army had heavier firepower. The rattle of gunfire and the deadly *zinngg* of arrows died out with the darkness. When the first rays of sun arose above the arid horizon the next morning, the Army men discovered that during the night the Paiute warriors, their women, and children had retreated into the bone-dry barrens of the Black Rock Desert, leaving their ancient village abandoned. If the Indians hadn't been outright beaten by the avenging army, at least they had been driven off.

The Nevada communities and the California trail were safe again. Once more Pony Express riders could carry the mail.

The Army regulars from the expedition later marched southeast to a point on the big bend of the Carson River about twenty-five miles east of Carson City, not far from the scene of the Williams Station massacre. There they built Fort Churchill, to protect the overland trail from further Indian attacks. Although more than a hundred years have passed since then, a few remnants of this isolated desert military outpost still are visible to those who turn off the highway to inspect them.

From the Sierras, hundreds of poles shaped from trees were hauled to the area and erected in a line across the desert from Virginia City to the new fort. A telegraph operator set up his batteries in one of the buildings. Soon he was "talking" across the mountains to California. Thus Fort Churchill, a remote and lonely desert outpost, became the easternmost point of the telegraph line from California. It was the first place where Pony Express riders bringing

A Pony Express rider hails construction men erecting the transcontinental telegraph, whose completion soon would eliminate the need for his services. Painting by W. H. Jackson, 1936.

the mail from the East could send their news speeding ahead by wire to San Francisco and the news-hungry Californians.

How strange it was for the soldiers in Army blue at Fort Churchill to learn from the Express riders about Manassas and other early battles in the Civil War that were being fought by their fellow soldiers 2,500 miles away, while they hung around their sun-baked desert huts watching for Indians! It was like two different worlds, held together by an easily broken thread.

Things were to change soon, however. An optimistic group of promoters from the Western Union Telegraph Company decided to make that thread of communication stronger and faster by building a telegraph line all the way across the gap between East and West—to make the dream of a transcontinental telegraph line come true, if they could.

CHAPTER 4

Stringing the Wire

When the first paddle-wheel steamboat of the 1861 spring season, churning up the Missouri River, drew near the raw frontier village of Omaha in Nebraska Territory, the captain yanked on the whistle cord. The bellow of the steam blast brought most of the few hundred inhabitants rushing to the muddy river bank to join the excitement. They milled around the landing, shouting to the boatmen. After dreary months of heavy snows and biting winds, they cheered this evidence that the bitter winter on the Plains had ended. Among the first to reach the scene was a young man named Charles Brown, fresh out of college, whose mission aboard the steamboat involved far more than curiosity.

Piled on deck was equipment for building the telegraph line that was to span the wilderness gap to California. Here was the material that would tie that distant state to the rest of the country by a single thread of wire at a critical moment in the country's history.

Brown's immediate task was to get the equipment off the boat and stacked on the bank.

"Pile them here," he ordered the gang of workmen hauling the large coils of wire and hogsheads of wooden insulators ashore. "The wagons will pick them up later." Other steamboats that followed brought more of the precious supplies upstream from St. Joseph, Missouri, until the bank was lined with high piles of the ingredients for building a telegraph line—large, thick, wicker-covered bottles, or carboys, filled with acid for the batteries, kegs of nails, boxes

Detail of turn-of-the-century map shows the Plains area west of Omaha. The number 1 indicates Fort Kearney; 2, Julesburg.

Edward Creighton

of sending and receiving instruments, shovels and pickaxes, iron bars, and dozens of axes. Without those axes the line could not be built. Thousands of poles necessary to carry the wire must be shaped from trees chopped along the route, and the disturbing fact was that on the Plains ahead were stretches of more than a hundred miles where no trees could be found.

Young Charley Brown had been hired as an assistant by Edward Creighton. That genial Irishman had the gigantic task of building the telegraph nearly a thousand miles across the prairie and through mountains to Salt Lake City, where Brigham Young presided over the Mormon church headquarters.

Far away back East, the armies of the North and the South were testing each other in the first battles of the Civil War that spring. Feelings about the war ran high even here on the distant frontier. So much had to be done in getting the telegraph building started, however, that concern about the fighting was somewhat subdued at the moment. Creighton with Brown's assistance worked at a frantic pace. Bosses and drivers must be hired for the wagon trains, hundreds of cattle and scores of heavy freight wagons purchased, crews of laborers assembled to cut poles and dig postholes.

One by one the heavy wagons, each pulled by six or eight oxen, were driven to the river front and loaded. The rolls of galvanized wire on crude reels were bulky, so heavy that some of the wagons when fully loaded sagged with nearly four tons of cargo.

"Put those wagons into line!" Creighton's booming voice rang out. Turning to his assistant, he ordered, "Charley, tell Matt Ragan to get that train of his rolling."

A covered wagon. Oxen were better than horses and mules for slow freight over a long haul.

Soon eight wagon trains were organized, each with a wagon master as its boss. At the crack of the drivers' long bullwhips, they headed west up the grade from the river, along a dirt trail through the scattered shacks of Omaha, and out into the Plains.

Behind lay months of planning and negotiating in Washington and San Francisco. Agitation for the transcontinental telegraph had grown louder and louder in California as the Civil War drew near. Would California stay loyal to the Union? Isolated as it was from the other states, political leaders in Washington and in California itself feared that the Golden State might declare independence or align itself with the South. The North needed California and the South wanted it.

Wise leaders in California sent a message to Washington: "The best thing you could do to keep our state with the North is to build the telegraph, so we can keep in touch with you."

It was this political need, more than a vision of big profits, that finally prompted Congress to attempt the job that many doubters thought couldn't be done. Too far, too much bad weather and difficult terrain, too much danger of Indian attacks—those were the arguments most frequently heard against the plan when Congress considered it in 1860. The outbreak of the Civil War was several months away, but its approach hung ominously over Congress that summer.

Southern members still sitting in the House and Senate lacked the votes to block the telegraph. Congress passed the historic Pacific Telegraph Bill on June 16, 1860, calling for competitive bids to determine which company would undertake the challenge. The law called for construction of a telegraph line from the Missouri River to San Francisco within two years, along any route the winning company selected. The United States government would subsidize the winning company with a payment of up to $40,000 a year for ten years in return for transmission of government messages over the line. If the government sent more than $40,000 in messages in any year, it would pay the company additional fees. Ownership of the land on which the line was to stand presented no problem. The government owned so many thousands of square miles of unoccupied lands in the West that the builder could take whatever he needed, free of charge.

When the time came for the Secretary of the Treasury to open the bids from the companies seeking to build the line, strange things began to happen. Western Union, which was developing into the most powerful telegraph company, bid $40,000, the maximum permitted under the law. Three other companies placed lower bids. One said it would do the job for a subsidy of only $25,000 a year. Since the law called for awarding the contract to the lowest responsible bidder, apparently Western Union had lost out. Then, mysteriously, one after the other of the lower bidders withdrew their offers. Nobody ever explained why they suddenly lost interest in building the line. Soon only Western Union was left in the competition, so it was awarded the government contract for its maximum bid. Behind-the-scenes deals between government and big business are nothing new, obviously

Actually, two short sections of the cross-country line already had been built when the government contract was signed in September, 1860, a 200-mile stretch at the eastern end from Omaha to Fort Kearney in Nebraska Territory and "Bee's Grapevine" across the Sierras to Fort Churchill.

When the original short local telegraph lines had been built in California during the Gold Rush years of the 1850's, they raised fears among the superstitious Mexican native residents of the new Pacific state. Seeing the poles with their crossarms, some Californians believed that the Yankees were fencing in the country to keep the devil out. Baffled by the mysteries of electricity, others got the idea that the wires were hollow and paper messages were being forced through them. These lines were put up by the California State Telegraph Company, which later struck a deal with Western Union to share in the transcontinental construction.

The deal, made early in 1861, was convenient for everyone concerned. A subsidiary of Western Union called the Pacific Telegraph Company would build the line as far west as Salt Lake City while a subsidiary of the California State Telegraph Company named the Overland Telegraph Company would build the line eastward from Fort Churchill, Nevada, to Salt Lake City. The two

Insulators on telegraph poles were favorite targets for gun-slinging, early-day Texas cowboys. In many places in the country, short, local telegraph lines were strung before the construction of the transcontinental line.

companies would share in the government subsidy and other income, 60 per cent to the eastern company and 40 per cent to the western one.

The eastern company had the greater distance to string the wire while the western one had rougher country. The jobs looked equally difficult, so the companies made a bet. The company whose line reached Salt Lake City first would get a bonus that would grow bigger every day until the other's line reached the Mormon city.

These financial arrangements meant less to jovial Ed Creighton than getting the job done. He was a builder, not a financier. His Irish immigrant parents in Ohio were so poor that he'd had to quit school after fifth grade. As a young man with a team of horses and a wagon he got a job hauling telegraph poles for a new line being built across southern Ohio and Indiana. Before long he was a superintendent of telegraph construction to New Orleans with forty crews working for him. If anyone could build the transcontinental line, Ed Creighton was the man. President Hiram Sibley of Western Union thought so and hired him as construction boss.

Sibley's orders were simple: "Build the line, and build it fast. We'll send you the supplies you need, but it's up to you to find the poles."

Hiram Sibley was the second president of Western Union. Later he spurred the Alaska Purchase.

To see for himself the task he faced, Creighton made a single-handed trip from Omaha to California in the winter of 1860. On this survey trip made at the request of Western Union officials in New York he lost his horse and equipment in the Platte River, rode a stagecoach part of the way, then switched to muleback. When he reached Carson City he was snowblind. His skin was burned from blowing sand and alkali dust. From there he followed the track over the Sierras to California in heavy snow. He finished his rugged journey weary but convinced that the telegraph line could be built along the trail he had followed. On Western Union's behalf he negotiated a construction agreement with the California State Telegraph Company.

After working out arrangements with the California telegraph people in San Francisco, Creighton returned to New York by the longer but easier ship route via the Isthmus of Panama. Finally

Illustration from an 1860 textbook shows a traveler resting at night on the hazardous trip across the Isthmus of Panama. Ed Creighton elected to return to New York by ship rather than repeat the cross-country winter trip he had just made. In the spring of 1861, he was back in Omaha ready to start construction of the telegraph line.

Ed Creighton's construction crews built the telegraph line through the hills near Fort Bridger, Wyoming, before the Union Pacific Railroad track was laid.

back in the ankle-deep mud of Omaha, he started his wagon trains on their trek to the West as the spring of 1861 grew warm.

In Creighton's expedition were four hundred men, five hundred oxen, and approximately a hundred freight wagons, divided into eight wagon trains. The grand plan for building the line called for dividing the work into several sections for simultaneous construction. Creighton was in over-all charge of the project as far west as Salt Lake City. One crew under his command was assigned to extend the line from its outpost at Fort Kearney two hundred miles to Julesburg, a tiny settlement in what is now the northeast corner of Colorado. Creighton assumed personal on-the-scene command of line construction from Julesburg to Fort Laramie, across Wyoming Territory through South Pass, past Jim Bridger's trading post on the Green River and down through the mountains into Salt Lake City. In turn the company from the Pacific Coast put one crew to work from Salt Lake City westward across the salt flats and past the Great Salt Lake, using supplies hauled by wagon from San Francisco. Its other gang worked eastward from Fort Churchill toward a link-up point.

Hardly had Creighton's cumbersome high-wheeled wagons creaked out onto the Plains from Omaha than they ran into serious trouble.

Mud. Black, sticky, almost bottomless mud—miles and miles of it. Winter's melting snows had turned the open ground into a quagmire. The wooden wheels sank to the axles. The straining oxen became mired, too. Despite angry shouts and the cracking of whips, at times the drivers could force the ox teams to move forward only two miles a day. It was as though the Plains were fighting back against these intruders.

At a place called Rawhide Bottoms, a stream had overflowed from the spring rains, forming a formidable obstacle. It had to be crossed. What the drivers would have given for a bridge such as the people back East took for granted! Creighton solved the problem by collecting many bales of rough, long grass called prairie hay. This was spread on the rutted tracks down to the stream bank at a point where the river seemed shallowest. The oxen were teamed up on one wagon at a time, sometimes as many as forty bellowing, restless animals yolked to a single vehicle. Shouting lustily, the driver sent the animals tugging, scrambling, and swimming to the far bank. Then they were swum back to move another wagon, until all were across.

Working in the wagon crews were a group of young Englishmen who had crossed the Atlantic to join the Mormons in Salt Lake City. The huge empty spaces of the prairies were strange to them; there was nothing like this back home. They knew nothing about oxen. Their attempts to yolk and handle the oft-times stubborn creatures caused hilarious laughter from the old frontier hands, already amused in their rough way at the manner in which the foreign "greenhorns" spoke and dressed.

Charley Brown had been sent ahead to Fort Kearney by stagecoach to arrange for the arrival of the first supply wagons. After two sleepless nights and a day of jolting, he arrived at the Army post and nearby settlement of adobe huts; together these made up one of the few "civilized" outposts on the road to California. The telegraph station, at the end of the wire from the East, was a low building of mud bricks and turf that sprouted clumps of grass on its roof and walls.

That evening Charley wrote a letter home. "The inhabitants of this little mud house city would not be worthy members of a church society," he reported. "They are preeminently hard cases and 'tough cusses.' In nearly every building in town there is a saloon with a small assortment of groceries as a side show. The chief occupation of the men is gambling, horse and cattle stealing and drinking whiskey." The frontier West, indeed!

Normally about six companies of United States Army men were stationed at Fort Kearney a short distance from the village, to protect the westbound wagon trains of emigrants from attacks by Indians. So heavy was the need for soldiers to fight for the North in the Civil War, however, that the garrison had been reduced. Only a single company of cavalry was posted there at the time. It was supposed to protect hundreds of miles of the trail, a physical impossibility if the Indians chose to attack. Would they? The question hung in everyone's mind.

CHAPTER 5

Charley Brown's Diary

From Fort Kearney, Charley Brown hitched a ride with a small wagon party to Julesburg, where construction of the telegraph line across the long, lonely miles of Wyoming Territory was to begin. Every night he wrote in his diary. Its pages give a fascinating picture of how the telegraph builders worked and lived.

June 19, 1861

Went to bed last night at about ten o'clock and for the first time this year tried to sleep outdoors. Made up my bed on the ground under a wagon. I found it impossible to sleep. My time and energies were given to fighting mosquitoes. The air was dense with them. I fled from them and took up quarters in Snydenham's ranch, who gave me permission to sleep on the ground floor in one of his rooms. In a short time I found I had made a great mistake in changing sleeping apartments. I not only had night birds of musical voice and long sharp poisonous bill, but worse and worse bed bugs in infinite numbers.

This I could not endure and again returned and made my bed under the same wagon. About midnight a good strong breeze came up, before which the winged pests disappeared and I had a good sleep.

June 20

Slept last night in Irish Tom's ranch on a buffalo robe spread on the ground. During the latter part of the night the lightning was very sharp and the thunder heavy. The lurid flashes of lightning and the deep heavy rolling thunder caused our horses to take fright and pull their lariat pins, they started east on a run. There was nothing for the boys to do but give immediate chase. They found them ten miles away.

An old print

Stopped at Smith's new ranch for nooning. There was an encampment of Indians at this place. About fifty of them had for shelter five tepees or wigwams. These tepees are made of poles covered with tanned buffalo hides. These poles are from twelve to fourteen feet in length and about three or four inches in thickness at one end and one or two inches at the other. These portable homes are not warm nor otherwise comfortable. The Indians sleep in them with feet to the center and head to the tent covering.

These Indians are the Ogalalla Sioux. The Sioux is the most powerful tribe in our country. These two tribes of natives are very splendid specimens of the human race physically and from general appearance mentally. This squad of Indians was moving toward Cottonwood Springs where they were collecting for the purpose of holding a war dance at which was to be exhibited the scalp of a Pawnee, taken in a late engagement.

June 21

At Morrow's met Matthers J. Ragan, who with his small number of men was cutting poles for the Pacific Telegraph Co. He had cut and hauled down to Morrow's about five hundred red cedar poles from what are known as Cottonwood canons. In a few days he was to haul them to Julesburg.

Emigrant train crosses the South Platte River in northeastern Colorado as Indians watch. The first telegraph line to California was strung across the Platte near this point.

June 23

Sunday! Today Sunday? Can it be and I here on the Plains over three hundred miles from Omaha? Sunday comes and goes with the unending regularity that marks all other days. All days remain the same. There is no change. No church bell rings the hour for Christian worship. Men continue their regular toil, not even ceasing in their swearing and drinking. The trains move lumberingly along as on other days. Here life is rough and wild, and how easily we glide into savage customs.

June 24

I gathered up my "traps" and moved westward, walking alone to Bouvier's ranch at what is known as Lower California Crossing, one of the places at which emigrants crossed the South Fork of the Platte River on their way to the Pacific. I stayed at this place until four p.m., when I walked to Baker ranch three miles west. Walking and carrying your bed and other traps on your back in the broiling sun is not a very pleasant pastime.

This country west from Plumb Creek to Denver is at present almost valueless for agricultural purposes. It can to some extent be utilized for grazing cattle and sheep if it were not for the wolves. The soil is dry and hard. There is no wood in this country, nor has there been any since I left Cottonwood, which is distant about eighty-five miles.

June 25

At about three p.m. arrived at Julesburg. At this point the telegraph line is to leave the South Platte Valley, crossing the river and thence for miles running up Lodge Pole Valley. Then crossing the [Continental] Divide, it is to follow the North Platte Valley. George Guy was here with several loads of poles from Cottonwood.

Julesburg is on the south side of the Platte and about two hundred miles west of Fort Kearney and four hundred from Omaha. It derives its name from a Frenchman named Jules, who for years has been an Indian trader. There are six buildings in this city, built of logs. One store or trading house, one dwelling house or stage station, and four sheds and barns. The stages for California diverge from the Denver route at this point and run to Fort Laramie, South Pass, Salt Lake, Virginia City, Sacramento and then to San Francisco. [This is approximately the route the telegraph line was to follow.]

June 27

With no special admiration for Indian character, as I have learned it on the Plains, I cannot refrain from saying that these wild men have endured much and long the aggressions and wrongs which the whites have extended to them before they, in their way, have sought to redress them. The wonder is, not that we have had Indian wars, but that we have not had more of them.

There is quite a large encampment of Cheyenne Indians here. They are on their way to White Man's Fork. These red men now are very friendly.

June 29

Ragan's teams came up today from Cottonwood and brought 399 red cedar poles. These will cover sixteen miles of the telegraph line. Today James N. Dimmock reached Julesburg on his way to Denver with three teams, loaded with bacon, flour, coffee, sugar, etc. Creighton bought his loads and hired him to work for us in constructing the line.

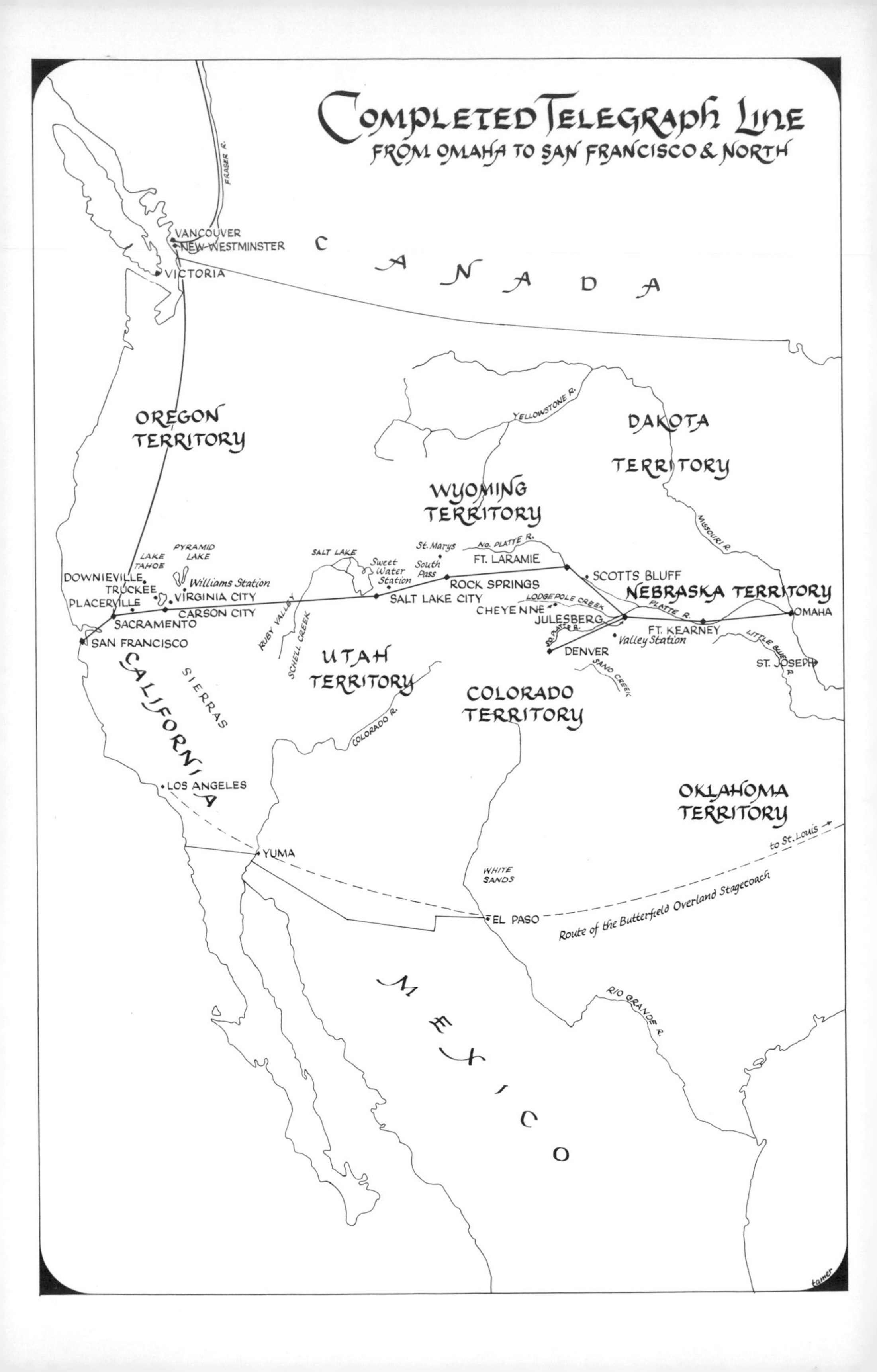
Completed Telegraph Line
FROM OMAHA TO SAN FRANCISCO & NORTH
FRASER R.
VANCOUVER
NEW WESTMINSTER
VICTORIA
CANADA
OREGON TERRITORY
YELLOWSTONE R.
DAKOTA TERRITORY
WYOMING TERRITORY
MISSOURI R.
PYRAMID LAKE
LAKE TAHOE
SALT LAKE
St. Marys
NO. PLATTE R.
FT. LARAMIE
Sweet Water Station
South Pass
SCOTTS BLUFF
DOWNIEVILLE
TRUCKEE
Williams Station
ROCK SPRINGS
NEBRASKA TERRITORY
PLACERVILLE
VIRGINIA CITY
SALT LAKE CITY
LODGEPOLE CREEK
CARSON CITY
CHEYENNE
PLATTE R.
OMAHA
SACRAMENTO
JULESBERG
FT. KEARNEY
SO. PLATTE R.
LITTLE BLUE R.
SAN FRANCISCO
RUBY VALLEY
SCHELL CREEK
Valley Station
DENVER
ST. JOSEPH
CALIFORNIA
SIERRAS
UTAH TERRITORY
SAND CREEK
COLORADO TERRITORY
COLORADO R.
LOS ANGELES
OKLAHOMA TERRITORY
to St. Louis
YUMA
WHITE SANDS
Route of the Butterfield Overland Stagecoach
EL PASO
MEXICO
RIO GRANDE R.
tamer

July 2

We commenced construction of the telegraph line today. The starting point was from the office established in the station house at Julesburg. We set fifteen poles and stretched the wire across the river. I helped Ed Creighton dig the first post hole. The wire was carried across the river on three poles—two tall poles, one on each bank of the river and one on an island in the river. These two tall ones were made from splicing two or more together.

July 3

Returning this morning to the station on the south side of the Platte, Mr. Edward Creighton, whom in the future I shall mention as "Ed"—the close, friendly name we all respectfully call him and which he likes—completed all the arrangements that were necessary to be made at Julesburg this afternoon, leaving young Mr. Reynolds, a telegraphic operator, in charge of the office established here.

We in the Concord buggy drawn by Mary and Jane crossed the river just as the sun was setting. We came very nearly upsetting two or three times on our way through the water. Once or twice the mules had to swim. We got our feet well wetted. Came to Hazard's camp on Pole Creek. The line of wire was stretched to this creek, a distance of about four miles from the Platte. An additional two miles of poles were set on which the wire was not strung.

July 4

The 4th day of July! I wonder if there will be many such for this nation as it now is. Can it preserve itself from disintegration? My prayer is that it may.

July 5

Ed and I started this morning for Mud Springs, distant from Julesburg sixty-four miles. All day long did we travel and see no vegetation over five feet in height. We crossed Pole Creek at what is known as the Upper Station about four p.m. From this point to Mud Springs our line of travel is nearly north over high rolling ground for twenty-seven miles. Not a drop of water is to be had on this route during the entire distance. We reached Mud Springs about seven p.m. and as expected found Jim Dimmock in camp here. His report on finding poles was very discouraging.

July 6

At about ten o'clock I left Dimmock's train to explore the canyons on our left and see if they contained material for telegraph poles. In prosecuting my search in the broiling sun, through ravine and over hills, I found myself suffering from thirst, and I soon turned from looking for poles to hunting for water.

On my way up into the canyons I saw a herd of antelope and their fright at seeing me sent them scurrying over the plains at full speed.

In the afternoon worked with the men in getting out poles. This was very slow and hard work. The "poles" were crooked, knotty and most of them good sized trees, dead, dry and brittle. We cut and piled eighteen. Getting such poles was very unsatisfactory, but we had to have them.

July 8

Took stage about noon for Lawrence Fork, about ten miles from Mud Springs. My object to this point was for the purpose of marking off the ground between this point and Mud Springs for distribution of the poles at or near equal distances apart.

We allowed from twenty-two to twenty-five poles per mile. If the ground was level twenty good poles were made to do the wire supporting. For the purpose of aiding the wagons in unloading the poles at the right place, I started at a point on the Fork and measured off seventy to seventy-three paces and then dug a small hole and made a mound of the earth excavated with a spade I carried. These mounds I made of sufficient size to attract the attention of the person on the wagon distributing the poles. About sundown brought up at the Springs pretty well worn out. The staking out or mounding out of ten miles was a big afternoon's work.

July 10

We are now settling into good hard work and the building of the line will progress from this time forward rapidly. The trains of Guy and Dimmock are at work in the bluffs and canyons on Pumpkin Creek about eight miles southwest of here. Hazard's men are over near Chimney Rock on the Lawrence Fork. Hibbard has charge of the construction squad of our men and is now working his way up Pole Creek. Ragan and Chrismen's trains are hauling poles from Cottonwood and Julesburg and distributing through Pole Creek Valley. Joe Creighton's train is somewhere east of Julesburg. Soon all the different divisions of our entire force will be at work on the construction of the line.

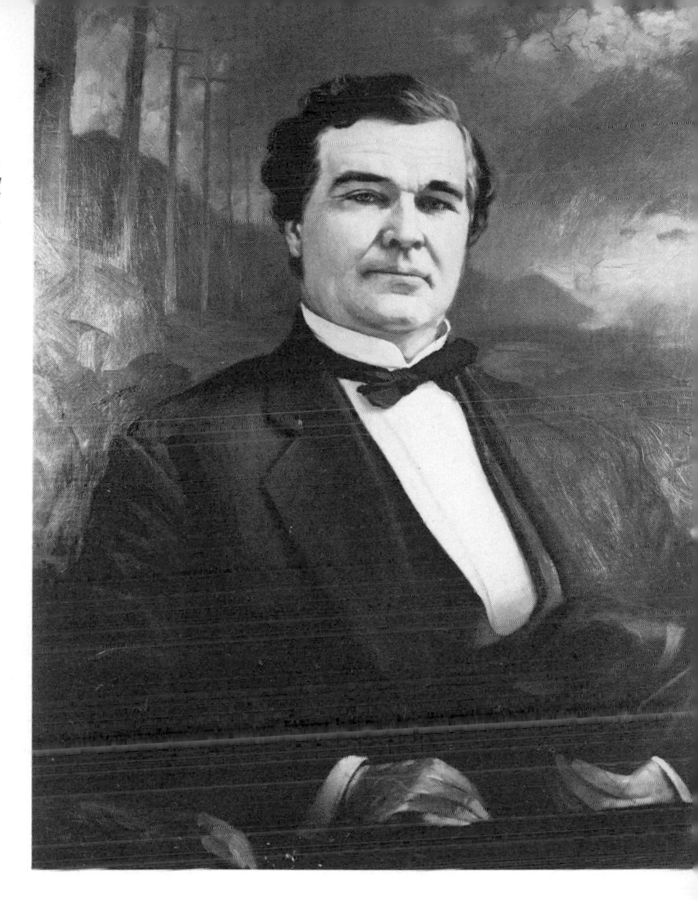

Portrait of Edward Creighton. The telegraph line he helped to build shows in the background.

Mr. Creighton estimates that we must average at least eight miles of the constructed line every day in the week and every day in the month. Counting the days we cannot work, to realize his estimate we must build each working day ten to twelve miles of the line. This I believe we will be able to do.

July 15

Started this morning with Cy and Christ, the hunter, for Dimmock's and Guy's camp to have them load and distribute the poles they had cut, and then move forward and find a new place from which to cut poles. As my companions had ponies to ride and I had none, I had to walk and so started in advance of them.

After about an hour's walk I discovered an antelope grazing not far away. I concluded to try and shoot him with my pistol. Putting a large rock between me and the harmless animal, I began to skulk and sneak

The prickly pear cactus

upon it. Unseen, I walked upon a bed of prickly pears of cacti and filled my moccasined feet full of their thorns. I forgot all about the antelope, and with a bound and some vigorous exclamations I was soon off the dangerous ground. Of course the antelope saw me and quickly made his escape. I removed the briars from my feet and moccasins as best I could.

The eastbound stage, which came in that night, brought the startling news that one of the Pacific Telegraph Company's men had drowned at Fort Laramie while attempting to cross the Platte River.

July 16

In the morning took breakfast at Chimney Rock station. The mosquitoes hung around and over the station in great swarms. In all my experience on the Plains I have never seen these bloody pests in such numbers. To protect the stage passengers from them during the morning meal there was a big smudge fire at the doorway, and in the dining room under the table two smaller ones. The smoke in the room was dense and almost suffocating, but still the bloodsuckers got in their work.

One can judge that these pestering insects were very thick, when I state that during the night they stampeded a small body of Indians who were in camp near the station. They were driven away and, pulling up their tepees, they moved over on an island in the North Platte River.

July 18

Started early this morning with the whole train for the foothills of Laramie Peak to cut telegraph poles. The train wended its way up the creek, keeping in the valley thereof. I left the train about nine o'clock and struck out into the hills as an exploring party. The train made only about eight miles before camping for dinner and cattle rest.

After dinner Jim Creighton and McCreary started on muleback to select a place to cut poles. Ed Creighton arrived with a span of mules and buggy. Ed is a steam engine of energy and has wonderful powers of endurance, and the enterprise he now has in hand virtually compels him to be ubiquitous. Jim and McCreary did not come into camp until very late. They had found poles in abundance at the base of Laramie Peak.

July 20

Called up this morning at light and directed to go in search of the mules, which had pulled up their picket or lariat pins and gone off. Saddling

An artist portrays a construction crew erecting the transcontinental telegraph line in the western mountains.

McCreary's mule, I started down the creek we came up and came upon them about four miles from the camp. After chasing them about three miles, I finally succeeded in driving them into a wild plum thicket and took them prisoners and returned in triumph to the camp.

July 22

At eight o'clock was on my way to Scott's Bluffs, through which I passed and reached Turjon's ranch eight miles east at three p.m. This was a sweltering hot day. During the afternoon H. Powell, one of Joe Creighton's men, came up to this ranch with two loads of wires and insulators and I helped him unload and store them away. From him I learned that the wire had been strung as far as Chimney Rock.

July 29

Commenced an early day's work and from last night's camp we distributed poles eastward through Scott's Bluffs to where the construction party had set poles and strung the wire. We rolled into camp at a late hour, weary and dusty. I drove the cook wagon. We met at Scott's Bluffs the construction gang of men who were engaged in digging the pole holes.

Billy, the cook, killed a mountain sheep and we had some cooked for dinner. As this was the first fresh meat we had had for a long time, the boys pitched in and did ample justice to the repast.

August 2

On the way up last night set a pole in place of one which had been destroyed by lightning.

August 5

Commenced my day's labor early in the morning, which was marking for the distribution of poles by Hazard. Went into camp at noon near the Beauvois ranch. Both Hibbard and Hazard's trains were resting together. Creighton connected a transmitter to the wires we had strung and sent messages of a friendly and business nature.

Quite an amusing incident took place at this encampment. Alongside of the Overland Emigrant Road and a short distance from Beauvois there was a fairly good-sized band of Cheyenne Indians in camp. After I returned from trading Creighton proposed that a few of us pay a visit to our neighbors.

The operator who traveled with Hibbard was taken with us. He carried with him an electric battery and other requisites for administering electrical

Charles M. Russell drew this sketch of Choctaw Indians moving when he was just a boy, long before he saw the West or an Indian. Courtesy Treasure Products, Inc., publisher of the C.M. Russell Boyhood Sketchbook, *Bozeman, Montana*

shocks. When we reached the camp we found a "squaw man" who could serve as interpreter between ourselves and the Indians. Creighton and Hibbard told them what we were employed at, that lightning ran along the wires, and talked and made them in a hazy way understand how dangerous it would be to interfere with our work.

Creighton then, practically, to illustrate what he was saying, proposed to give them a shock from the battery. This being arranged, Hibbard generated a very powerful electrical charge. Joining hands with the Indians, squaw and buck, Hibbard "let her go," and there was prancing then and there among the Indians. They did some talking and looked upon Mr. Creighton as a "big medicine man."

When Hibbard commenced stringing the wire that afternoon it so happened that he had to run it directly over the tepees of this camp of Indians. In hot haste the squaws and also the bucks tore down their lodges and decamped.

When we quit work that night our telegraph line was complete to three miles west of Fort Laramie.

CHAPTER 6

Hello, San Francisco

A construction worker rode into Ed Creighton's telegraph crew camp in Wyoming Territory and reported excitedly, "They're knocking the poles down, Ed!"

"Who is?" the superintendent replied. "The Indians?"

"No, not them. The buffalo. Dozens of poles are flat on the ground."

The great shaggy, broad-shouldered animals that roamed the prairies in massive herds had indeed become a menace to construction of the telegraph. They discovered that these posts sticking out of the ground on the treeless prairie made wonderful back-scratchers. Several buffalo would congregate around a pole, taking turns rubbing against it with the bison's equivalent of a contented smile. Within hours their heavy sidewise nudging would force the pole out of the ground and snap it in two.

Tellers of tall tales soon claimed that within a few hours after word of these wonderful new back-scratchers spread among the buffalo herds, waiting lines of three hundred buffalo had formed at every telegraph pole between Fort Laramie and Omaha. When a lone buffalo was seen trotting across the Plains, the story was that he had heard of a vacant pole "somewhere this side of Omaha."

Creighton thought he had a solution. He told a work crew, "Hammer long spikes into the poles. That should keep them away."

The workmen carried out his instructions along several miles of poles in an area known as one of the traditional trails the buffalo followed on their treks north in summer and south in winter. A

Charles M. Russell drew this sketch of buffalo scratching their backs on the spikes of the telegraph poles. Courtesy Treasure Products, Inc., publisher of the C.M. Russell Boyhood Sketchbook, *Bozeman, Montana*

few days later the wagon-train boss reported to Creighton in disgust.

"It doesn't work. The buffalo like the poles better with spikes. Those long points dig through their matted hair and scratch their hides where it itches." The spikes had to be removed.

While Creighton's crews were working their way through buffalo country, hundreds of miles to the west the construction parties of the California-based Overland Telegraph Company were hanging wire across the tawny deserts of Nevada and Utah. One party worked eastward from the remote outpost at Fort Churchill. The other party moved westward from Salt Lake City, intending to connect somewhere in the desert and thus complete the western half of the system, from California to Salt Lake City.

Remembering the Indian uprising of the previous year that had disrupted the Pony Express and killed so many Nevada settlers, the construction crews were on the lookout for attacks. However, the Indians of the Nevada desert allowed the work to go ahead uninterrupted. The telegraph people made gestures of goodwill toward the Indians so successfully that some of the natives, especially the Shoshones, became friendly and showed much curiosity about the mysterious lightning line the white men were putting up.

Chief of the Shoshones was Sho-kup. Although he found the ways of the white men strange, he wanted to know more about them. He rode in from the barren Ruby Valley to Carson City on horseback and was given a return ride close to his home on the stagecoach. His hosts told about the wonders of San Francisco, far away over the mountains on the Pacific Coast, and promised to take him there on a visit. One day Sho-kup rode into Reese's River, a tiny settlement in central Nevada which the telegraph line from the west had reached. He announced that he wanted to send a message of goodwill to the big chief of the telegraph company in San Francisco.

John Yontz, the operator at Reese's River, agreed that this was a splendid idea, but it presented difficulties. His own knowledge of the Shoshone language was skimpy at best. Sho-kup's knowledge of English was equally bad. Getting Sho-kup's ideas harnessed into the dots and dashes of Morse Code was tricky business. Yontz did it, after a fashion. Over the wires to San Francisco went this message:

To Vice-President Telegraph

I saw big Telegraph Chief, Carpentier, on stage. Had shake hands with him. I like him and like telegraph. My Indians all shake hands with white man. My Indians shall not hurt telegraph line. White man good, then Indian good. I count my Indians now. Have about 5,000. About six weeks I go San Francisco to see steamboat ship and big water. All telegraph men treat me well. Street good man. Little Johnny [Yontz] telegraph good man. Hubbard and McDonald good man. Overland Mail good. Buckley good man. Wash. Jacobs good man, Jim Jacobs good man. Today me go to Ruby Valley.

SHO - KUP
Chief of the Shoshones

Although the Indians were peaceful, the Civil War raging back East created problems for the telegraph project. The plan had been to relay telegraph messages from the West at Omaha to eastern cities, over the existing telegraph line that ran from Omaha across Missouri to St. Louis. But Missouri had strong southern sympathies. So during the summer of 1861, the governor ordered the Omaha-St. Louis telegraph line torn down in order to handicap the transcontinental line, owned as it was by Northerners. Emergency construction of a new telegraph line became necessary, from Omaha to Chicago across Iowa, through land that was safely in northern control.

A heavy load of wire for the cross-country line was to be shipped up the Missouri River to Omaha, then hauled by freight wagon across the prairies and through the mountains to the Salt Lake City area. This shipment was blocked in Missouri by southern sympathizers. A new load of the essential material had to be shipped from New York around South America to California and then hauled east across the Sierras, a detour of many thousands of miles.

Despite these troubles, construction of the telegraph moved ahead at a steady pace during late summer and early fall of 1861. As the wires were stretched from the east and from the west, the gap between them grew smaller. When the Pony Express rider in each direction reached the outpost telegraph station, he handed his news to the operator to be wired ahead, before riding on as fast as ever to deliver the letters he carried. Many of these were private business not intended to be put on the telegraph.

Thus, typically, *The New York Times* published a report from California on September 22, 1861:

SWEET WATER STATION
Pacific Telegraph, 34 miles east
of Salt Lake City, Saturday
Sept. 2.

The Pony Express passed here at 8 o'clock this morning, bringing the following from the press.

There followed a series of news items from San Francisco about the arrival of ships and the California election results. The writer in San Francisco lamented that the local papers had been short of news from the East, because back in Missouri the bridges of the Hannibal and St. Joseph Railroad had been burned in a Civil War raid, causing a stoppage of news and letters that normally would be carried to St. Joseph, Missouri, by the eastern telegraph and railroad and handed over to the Pony Express there. So flimsy

Eight years after the telegraph crews strung the line through the mountains, an excursion party posed beneath the 1000 Mile Tree at a point almost exactly 1000 miles west of Omaha.

was the line of communication across the Great Plains and mountains that a small war incident two thousand miles away left the entire Pacific Coast population without news from the East for many days.

Slowly Creighton's work crews advanced the wire toward Salt Lake City. Always they struggled to find enough poles on which to hang the galvanized iron thread. As the line reached South Pass in central Wyoming, the scene of a gold rush a short time earlier, Creighton convinced the last of the disappointed prospectors to quit their hunt for gold and work for him in a different kind of hunt, for trees big enough to carry his wire. Considering the poor luck most of the gold seekers had suffered, probably he didn't need to exercise much of his sales charm to recruit them. The wages he paid were small but were in cash and food, which was much better than their get-rich-quick dreams that had failed.

To the crews working toward Salt Lake City, that Mormon community in the valley of the Great Salt Lake became a more glittering objective as each additional week passed in the lonely open spaces. With a population of ten thousand, it was by far the largest community from the Missouri River to California. No other settlement along the entire stretch of the route had more than one-tenth that population.

Fascinating, often grossly exaggerated stories of life in the Mormon city circulated among the men. Led by Brigham Young, the colony of Mormons that had been driven out of Nauvoo, Illinois, by religious persecution because of their unusual beliefs had made the long trek along this same route in wagons or on foot in 1847, searching for a home in the wilderness. One group of Mormon settlers who followed a little later pushed wheelbarrows the entire distance. As Brigham Young led the original party down through a pass in the Wasatch Mountains and saw the Salt Lake Valley spread out before them, he proclaimed, "This is the place." Now, fourteen years later, under Young's vigorous, stern leadership, the Mormons had irrigated the land, grown trees, vineyards, and crops, and built a prosperous community.

Most of the gossip among the telegraph crews centered around the fact that the Mormons in Salt Lake City practiced polygamy. According to their beliefs, it was not only permissible but desirable for a man to have several wives. Brigham Young himself was "sealed" to more than a score of women. He lived in an attractive home with his first wife, while ten more wives and their children lived in small apartments in a building nearby. Naturally such arrangements, which were contrary to religious principles and the law elsewhere in the United States, created wild stories about the goings-on in Salt Lake City. No wonder the telegraph crews were so curious to reach the city and see for themselves! What they found disappointed many of them. The practice of polygamy was carried on quietly under religious auspices, and public life in the city was strict, even somber.

Creighton realized that it would be good politics for the builders of the telegraph line to be on friendly terms with Young, so he hired one of Young's sons who was in the lumber business to supply

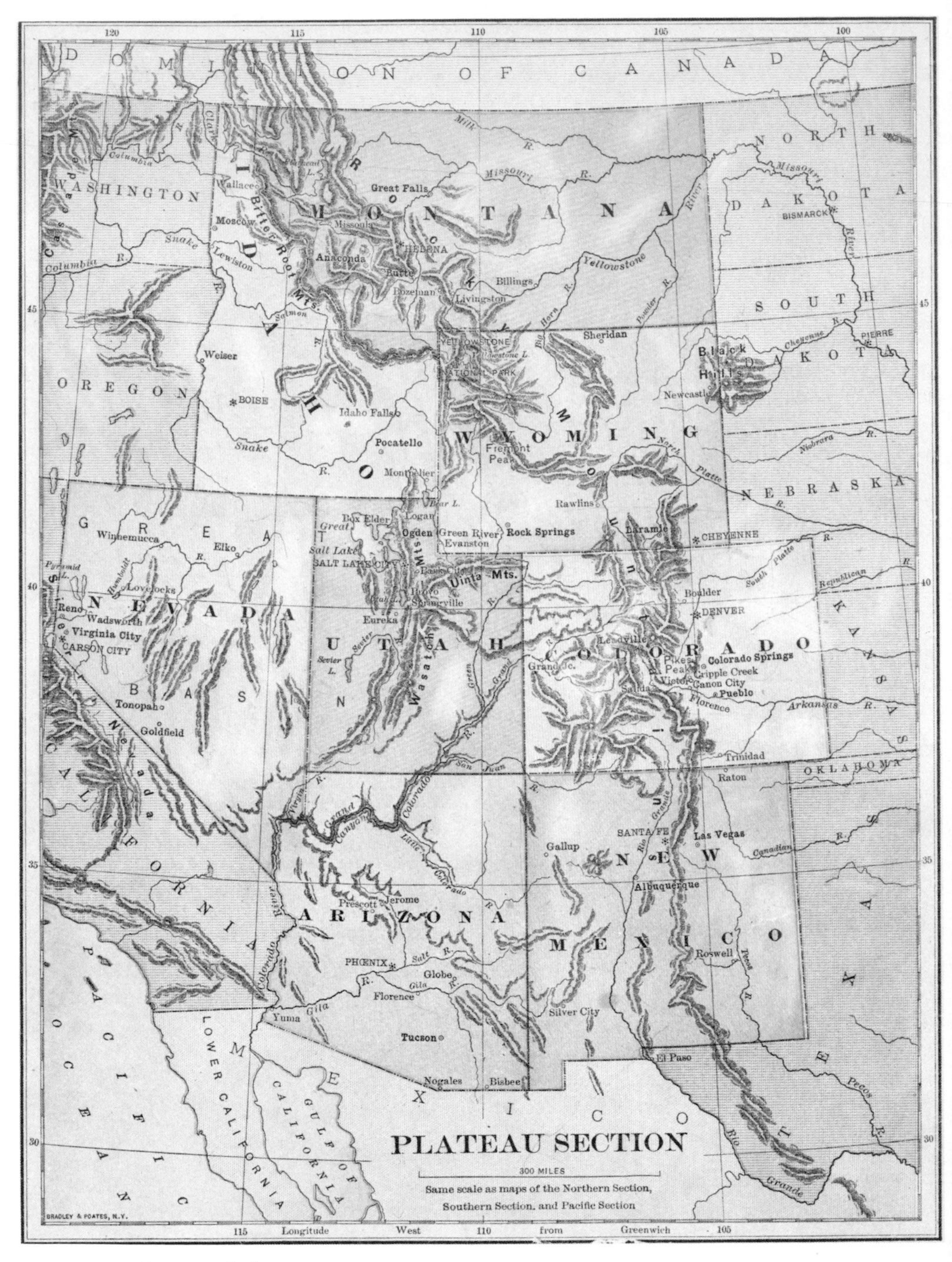

A turn-of-the-century map

poles. Although they had agreed on the price, the son complained later that the price was too low and he couldn't make enough money on the venture. Reluctantly, but hesitating to offend the powerful Mormon leader's son, Creighton agreed to pay a higher price.

A few days later Creighton received a summons from Brigham himself. After welcoming Creighton to his plain, neat office, the bearded "Lion of the Lord," as his followers called him, said to the telegraph builder, "Please let me see the contracts with my son."

Creighton handed them to him. Young studied both, crumbled the new one in his fist, and threw it into the fire.

"The poles will be furnished by my son in accordance with the terms of the original contract," he said.

The superintendent of telegraph construction west of Salt Lake City, James Gamble, also had difficulty with a Mormon businessman, who failed to supply the poles he had promised. When Gamble complained to Brigham Young, Young denounced the Mormon contractor from the pulpit and ordered that the contract be filled. It was.

By early October, Gamble's crews to the west and Creighton's to the east were drawing close enough to a linkup for both groups

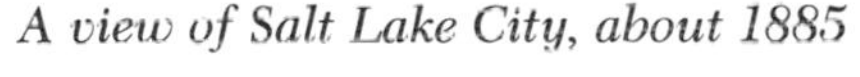

A view of Salt Lake City, about 1885

to realize that the race for completion of the wire to Salt Lake City would be close. At this point, Gamble's crew, working across eastern Nevada in a desolate stretch between Ruby Valley and Schell Creek, ran completely out of poles. All progress stopped.

Gamble sent parties of scouts into the mountains on a search for trees big enough to cut into poles. On a mountain-top fifteen miles from Egan Canyon they found a large grove of trees that was just what was needed. Greatly encouraged, Gamble sent a train of twenty heavy wagons into the mountains with a crew to cut the trees and bring them down. Already, late fall had come to the area. The air had a bite. Low clouds hung over the land. Winter with its threat of early snow was close at hand, and Gamble knew how little time he had.

Several days later a man returned from the wagon train with bad news. He told Gamble, "The men have refused to go up the mountain this late in the fall because they might be trapped by a snowstorm."

Gamble was furious. He and his assistants climbed aboard the first stagecoach they could find that was going to the area where the wagon train stood idle. Facing the rebellious men, he told them, "If you don't go up that mountain to cut those trees, you'll forfeit all your pay!"

The men mumbled among themselves. A spokesman challenged the boss: "We'll take a chance and go up there if you will come with us."

Gamble thought of all the duties as superintendent that he would be leaving undone if he went. But what did all that matter if there were no poles available? Unless those poles could be brought out quickly, he knew, winter would close in and construction work would be called off until next spring.

"Fair enough," he replied. "I'll go if you will."

So the wagons rumbled laboriously up the barren mountainside, winding their way through the rocky terrain until they reached the grove at the top. In two strenuous days of work that ran until darkness no longer permitted the use of an ax, with an eye turned always toward the heavy clouds banking in the west, the men chopped enough trees to fill the twenty wagons. That was sufficient

Old print shows logging operations in the mountains.

to finish the gap in the western line. Valuable days had been lost, however. Gamble feared with reason that Creighton's men advancing through the Wasatch Mountains that guard the eastern approach to Salt Lake City would arrive at the Mormon capital first.

Creighton's wire men reached Salt Lake City on the morning of Friday, October 18. They fastened the rolls of wire to poles already placed by the advance crew, moving past the farms on the city's outskirts, then along the dirt streets past rows of sun-dried adobe brick houses. Reaching Main Street, they pulled the wire into a small store building that had been designated as the telegraph office.

An operator accompanying the party set up his batteries in the bare room and placed his sending and receiving apparatus on a table. The wire was connected. Experimentally he tapped out a few dots and dashes. In return he received a test signal across hundreds of miles of wire from the operator in Omaha. Someone noted that the time was 1:00 P.M. Suddenly Salt Lake City, so remote from other cities, found itself in instant communication with the outer world.

Naturally, the first message to be sent from Salt Lake City was from Brigham Young, addressed to Jeptha H. Wade, president of the Pacific Telegraph Company in Cleveland, Ohio. After praising the project, Young added words that were happy news to the North:

Telegraph office on Main Street, Salt Lake City, into which wires were run from east and west in October, 1861, completing the transcontinental telegraph. Note the telegraph poles in foreground with tree limbs chopped off.

"Utah has not seceded, but is firm for the Constitution and laws of our once happy country"

Soon the receiving instrument clicked a staccato message from Washington. It was signed A. Lincoln. The President's words showed how important he considered the telegraph in holding the North together: "The completion of the telegraph to Great Salt Lake City is auspicious of the stability and union of the Republic and the government reciprocates your congratulations."

Six more days passed before Gamble's men completed the wire from the west. At last the job was done. Soon this wire came alive with chattering dots and dashes from faraway San Francisco. In that little room one instrument "talked" from California while another "spoke" words sent from New York. The operator received the messages from one wire and relayed them onto the other. Finally the United States was tied together by the electric telegraph from coast to coast. It hardly seemed believable.

In a telegram to the mayor of New York, the mayor of San

Francisco said, "The Pacific to the Atlantic sends greetings. And may both oceans be dry before a foot of all the land that lies between them belong to another than our united country."

An exuberant editorial in *The New York Times* the next morning caught the spirit of the occasion. "It is with almost an electric thrill that one reads the words of greeting yesterday flashed instantaneously over the wires from California. The magnificent idea of joining the Atlantic with the Pacific by the magnetic wire is today a realized fact. New York, Queen of the Atlantic, and San Francisco, Queen of the Pacific, are now united by the noblest symbol of our modern civilization."

While readers marveled at seeing dispatches from the opposite sides of the country published in the morning papers, the men who built the line cashed in on their gamble. They had taken a huge risk against the wilderness and the weather and had won. The line was completed nine months earlier than their contract with the government had required. In their joy and hunger for money, they

The original Western Union office in New York, maintained and operated from 1856 to 1930

conveniently forgot another section of their contract with the government, that they would charge no more than $3.00 for a ten-word message. They charged a dollar a word, a rate so high that only the rich could use the transcontinental line. So great was the novelty, however, that business boomed. Western Union soon accumulated great profits. Ed Creighton took part of his pay for building the line to Salt Lake City in Western Union stock, which grew to be worth more than a million dollars. With this stake, he developed a large fortune in business and banking ventures. His name is commemorated in Creighton University at Omaha, which he founded.

Even in their moment of triumph, the ambitious telegraph builders were not content. Having conquered the continent, they wanted to conquer the world with their strands of galvanized wire. Their next goal was to extend the telegraph line from San Francisco to Russia and thence to western Europe. The audacity of their plan was tremendous. So, too, were the profits they visualized if they could control the global telegraph business.

CHAPTER 7

The Indians Attack

While the telegraph promoters dreamed of wires stretching on around the world from San Francisco, the men who operated the new transcontinental telegraph were having trouble keeping it working. The wire stretching nearly two thousand miles was subject to many ailments and often was out of operation for hours, or even days, at a time.

Weather was the worst bugaboo at first. Lightning bolts hit the wires. In the first winter the line was open, the worst blizzards in many years blustered across the Plains, blowing snow almost horizontally in white clouds so thick that a man could see only a few feet. Winds smashed down the wire and snowdrifts buried the knobby, scrawny poles. Linemen riding head down through the storms, hunting the breaks in the wire, suffered extreme exposure.

Out in California, a great flood swamped the Sacramento Valley in the winter of 1861-62, a few months after the line was completed. This put the telegraph out of action over a wide area. Two repairmen who set out from Stockton in a skiff rigged with a sail, headed for Sacramento, found the water up close to the wire in many places. At one point they sailed right over the top of the submerged wires. Short circuits knocked out the system for days.

Some businessmen who didn't understand how delicate the telegraph circuit was grew suspicious that the repeated interruptions of service to the East from San Francisco were part of a conspiracy. Sending messages about the price of gold on the San Francisco

Exchange to traders in the eastern markets became one of the most important commercial functions of the line. So when the service broke down at crucial times and the San Francisco operator said he couldn't raise an answer from Salt Lake City, angry gold traders claimed that the breakdown was intentional, to withhold information on current prices so other speculators could make a killing. Who knows if any truth existed in this suspicion? It never was proven.

The greatest peril of all those envisioned by the builders of the telegraph, Indian raids, failed to materialize while the line was being built. While the project was being planned, stories circulated about "hordes of savages" on the plains and desert, waiting to swoop down on the construction crews and murder them. Tales of Indian attacks on emigrant wagon trains, and of the raids on Pony Express stations in Nevada a few months earlier, were spun around the campfires of the telegraph crews, never growing smaller in the telling.

The work gangs were heavily armed. Whenever a larger than normal group of Indians gathered around the telegraph camps, as they sometimes did when on the trail to their tribal meetings, uneasiness surged through the white men. Since the natives and the work crews knew little of each other's language, misunderstandings often arose.

Perhaps the Indians were so fascinated by the strange "lightning line" that they were more interested in watching it being built than in destroying it. Creighton, Gamble, and the other bosses made friendly gestures toward the Indians and ordered their men to treat the tribes fairly. Indian raids were the last thing they wanted to face; their job was hard enough as it was.

The principal reason the Indians left the telegraph builders alone, however, was that they did not comprehend the significance of the line. Their nomadic life of hunting and fishing had not prepared them for dealing with the kind of civilization the telegraph represented. At first, they did not realize that the telegraph was the white man's tool for reducing great distances and beckoning more settlers to follow the emigrant trail—settlers who would build towns on the Indians' hunting grounds, kill the buffalo herds, turn the

Old print shows Indian hunting buffalo.

plains into farms, and force the Indians into more remote, less habitable regions.

Always the builders tried to impress the wonders of the wire upon the Indians. The natives' knowledge of electricity consisted of nothing more than watching lightning bolts flash from the heavens. Smoke signals and drums were the extent of their ability to communicate through the air. One method the builders used was to assemble a group of tribal chiefs who were asked to agree upon a story among themselves and a sequel to the story, without telling the whites what it was. The chiefs separated to telegraph stations miles apart. One group told the story to a telegraph operator at one station; while they watched in awe, he clicked off their tale in dots and dashes. These were taken down by the receiving operator, who repeated the story to the second group. Deeply impressed that the wire knew exactly the story they had made up, they told their sequel to the operator, who transmitted it to the original group. They, too, were amazed. Indeed, the white man had a marvelous machine!

Except for an occasional Indian raid that caused little damage, the telegraph was safe until 1864. In that year, the Civil War was at its peak of ferocity. Because the Union armies needed every man they could muster, the frontier forces along the emigrant road and the telegraph line were guarded only by a few soldiers. The Indians sensed the white man's weakness. Angry at what their chiefs considered mistreatment by the whites, the Indians opened a general attack along hundreds of miles of the route across Nebraska, Colorado, and Wyoming. Among the thinly spread garrisons a rumor spread that the Indians were urged on—indeed, even led—by agents of

Each wagon train headed west was a threat to the survival of the buffalo and therefore the Indian.

the Confederacy. These Southerners supposedly sought to sabotage the North's lifeline to its loyal supporters in Salt Lake City and California. Never was proof of this suspicion established.

In their raids that year against the widely scattered ranches and stations, the Indians killed many women and children. By the white man's military standards of the period, this was an unforgiveable outrage. War was men's business. Indian traditions were different. They had their own code of conduct toward women and children which, while strict in many ways, did not draw such a line between the sexes in war. Now, a hundred-odd years later, the white man's methods of warmaking, with mass saturation bombing that draws no distinction among men, women, and children, resemble those of the Indians which he found so revolting in the 1860's.

Seeking to punish the Indians for these killings, about nine hundred American soldiers swept down on a camp of Cheyenne and Arapahoe along Sand Creek in Colorado on November 29, 1864. The soldiers struck at dawn while the Indian camp was barely stirring. They hammered the settlement with howitzers. Outside the lodge of Black Kettle, one of the chiefs, an American flag flew from the long lodgepole. This was an indication that the Indians considered themselves under American protection at the moment. The cavalry and infantry charged into the camp, chasing the Indians up the almost dry creek bed. The result resembled a massacre. Indian women and children perished alongside the men, many of them scalped by the soldiers in the same manner that the Indians had assaulted victims in their raids. Some reports said that three hundred Indians died that day; probably the true number was lower, but the Indians were stirred to even greater ferocity against the white settlements.

Among their main targets were telegraph stations and the telegraph line itself; the stations because they were isolated and the line because it was easy to attack.

Near South Pass in Wyoming, five men formed the little garrison of St. Mary's stage station and telegraph office early in 1865. Word flashed along the wire that the Indians were on the warpath, so the men were on the alert.

"Look!" one of them cried. "They're coming."

Indeed they were—150 Cheyenne and Arapahoe braves, galloping across the open rocky slope toward the stone station building. A stand-up fight was hopeless, the five men realized. They fled to an abandoned dry well some yards away. All they could save from the station was the telegraph operator's instruments and a coil of wire. Crouching in the hole, they watched the Indians race around the station and cut four hundred yards of the telegraph line.

The Indians shot their guns and fired arrows into the building in an effort to drive out the men they believed to be hiding inside. Nobody appeared. Puzzled, the Indians dismounted and ran into the station with tomahawks raised, only to find it empty. They looted its contents and set it afire. From their hide-out, the five white men watched in apprehension, not only from fear of being discovered but because they knew that ammunition was stored in the blazing building. Soon it happened; the ammunition blew up with a roar. Terrified, the Indians rounded up the station's cattle and rode away. Late that night, cautious because the Indians might still be nearby, the five men climbed from the well and slipped away in the darkness. They followed the line of poles west until they believed themselves at a safe distance. The operator climbed a pole, connected his instruments, and tapped out a message to Fort Bridger relating their escape.

Soon the wire was strung again and telegraph service east from Salt Lake City resumed. Western Union customers in the East continued to grumble: "That miserable telegraph has broken down again! We can't get any news from San Francisco. Why can't they keep the thing in order?" If only they had known what really happened when the Eastern operator reported, "Wire trouble between Ft. Kearney and Salt Lake City." Trouble, indeed, of a deadly variety!

When the Indians couldn't raid a station, they resorted to tearing out parts of the line in isolated areas. Burning the poles proved to be too slow, so they developed a trick. Throwing a rope over the wire and holding both ends of the loop, they galloped off until the wire was jerked free from its insulators on the pole crossarms and snapped. Night after night, soldiers working as repairmen muffled the hoofs of their horses and rode along the line in the dark, to restore the flow of electricity without alerting their enemies.

On the unbroken plains of Nebraska, the Cheyenne, Arapahoe, and Brulé Sioux almost simultaneously attacked emigrant trains, ranches, and telegraph stations along the Platte River. Plum Creek was hit first. The operator there managed to flash a warning along the wire, saving lives. When the attacks spread to the valley of the Little Blue River, where no telegraph line existed to carry the alert, the settlers suffered many deaths.

Where the city of Grand Island, Nebraska, stands today, recent German immigrants led by William Stolley operated the O.K. Store and telegraph station. As news of the murderous attacks and scalpings a few miles away reached the tiny settlement, many of the whites dashed frantically for the protection of Fort Kearney, driving their livestock before them. This proved to be a fatal blunder for some who were caught on the trail by Indian bands and killed. Led by Stolley, the Germans decided to sit tight and get ready for an attack. They built a mud wall and dug a ditch around the store. They packed three thick layers of sod, ripped from the prairie, around Stolley's house nearby until it was a dirt-walled fortress twenty-four feet in diameter with loopholes through which the settlers could shoot. They excavated an underground stable twelve feet wide and eight feet long where they sheltered their horses.

They were telling the Indians, "Come and get us if you dare." The Indians didn't. Hit-and-run raids with flying arrows, blood-curdling shouts, and pounding hoofs were their method; when the settlers were able to prepare themselves, the raiders stayed away. Siege warfare wasn't their style.

The telegraph operator at the O.K. Store got a message through to Fort Kearney reporting their situation. Back came word that the United States Army would get a supply of arms to the entrenched Germans. A few days later, as the situation quieted down, a detachment of soldiers appeared, but the arms they brought didn't amount to much.

Stolley wrote to his brother: "Instead of the sixty rifles and 1,000 cartridges promised by the governor we finally received sixteen old, bent blunderbusses on some of which screws and locks were missing; the cartridges failed to appear altogether."

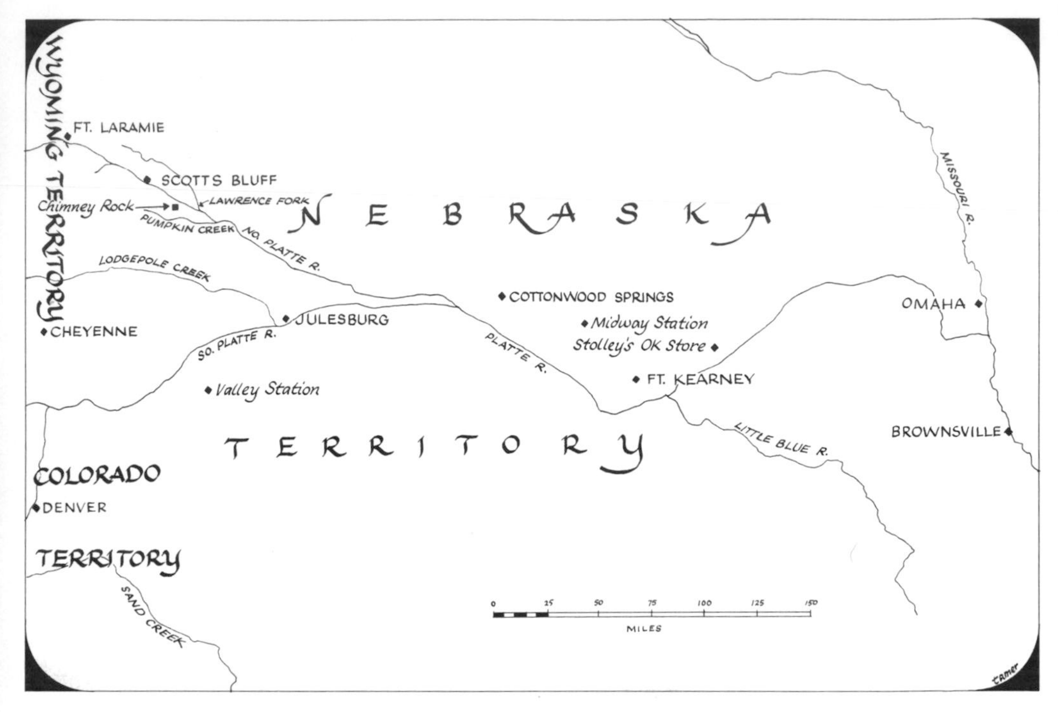

Stolley's OK Store, shown on the map, is the site of today's city of Grand Island, Nebraska.

Even so, when the Indian raids along that portion of the emigrant road died down, the O.K. Store and telegraph station, along with Stolley's house, remained intact behind their massive walls of mud, a lesson in what preparedness can do.

Farther west at Julesburg, where Ed Creighton, his assistant Charley Brown, and their construction gang had started building the line to Salt Lake City less than four years earlier, the Indians struck an especially crippling blow to the telegraph. On a bitter cold morning in February, 1865, fifteen hundred Indian warriors galloped screaming into the settlement on the treeless plain. Julesburg wasn't much larger than it had been when Creighton had his headquarters there; just a half-dozen scattered buildings, including the telegraph station and a blacksmith shop. A few hundred yards from the station was the military post, the only fortified building in the area.

Seeing the Indians coming, about fifty citizens ran to join the small garrison of soldiers in the fortified shelter. The soldiers had one howitzer and their rifles. Outnumbered more than ten to one, their only course was to remain within the fort and let the attackers have the run of the settlement.

The Indians came out of the north. They crossed the Platte River at the place where Creighton had planted his first telegraph poles. Thick ice covered the broad, shallow stream. An advance party of Indians spread sand on the ice, a path six feet wide, so the war party could cross without having their horses fall. Racing among the buildings, they set the telegraph office afire, put torches to the haystacks where last summer's crop had been stored as fodder, and rounded up the grazing cattle which provided much of Julesburg's food supply. Smoke from the burning hay hanging close to the ground partially obscured the view. Part of the Indian force circled the fortified, sod-banked military post, showering arrows into it and exchanging gunfire with the men inside. Other parties set out to destroy the telegraph line. They pulled down the poles and wire for ten miles to the northwest of Julesburg, a short distance to the east, and for forty miles along the branch telegraph line that ran southwest to Denver and the mining camps near it. Julesburg was cut off from the world.

Not aware that the raid was in progress, a party of eleven United States Army men escorting four civilians and a stagecoach rode toward Julesburg from the east, up the south bank of the Platte. They were coming to reinforce the garrison. With them they had a horse-drawn howitzer. It was an odd procession: Captain Nicklaus J. O'Brien and Lieutenant Eugene F. Ware in front, the stagecoach and its passengers, the piece of artillery with its five-man crew, then four other cavalrymen.

Julesburg was hidden from their view by a promontory of land jutting close to the river. A column of smoke rose behind this bluff. Across the river they could see Indians driving herds of cattle and horses. Something was wrong, obviously.

The party halted. "Inspect the gun and make it ready to fire," Captain O'Brien ordered.

"It won't fire, sir," a gunner reported. "The priming wire is lost."

To the party's dismay, the howitzer's priming wire had disappeared, apparently jolted loose along the trail. Without it the artillery piece could not be set off. In those old-fashioned weapons, the explosive charges were in thick flannel bags that were rammed down the barrel. The priming wire had to be jammed into the covering to make an opening through which the fire could reach the explosives and discharge them. Here was a crisis, indeed! An Indian attack ahead, and the one available artillery piece out of action.

Frontier ingenuity, born of necessity, saved the situation. Nearby ran the telegraph line. With an artillery hatchet, the soldiers chopped down a pole, allowing the wire to sag near the ground. The men cut the wire. They fastened one loose end to the back of the stagecoach and drove its four-horse team forward until the wire was stretched taut. Then they did the same thing with the other wire end.

Lieutenant Ware commanded, "Now bring the two ends together and see how much they've been stretched." They overlapped by two feet. "Cut off the extra length and connect the ends again."

When this was done, the telegraph wire was intact once more and the soldiers had a priming wire—two feet of stretched-thin wire that had been carrying the dots and dashes of cross-country telegraph messages only a few hours earlier.

Lieutenant Ware climbed the high point of land with his field glass to see what lay beyond. At the top he kneeled behind a bush and peeked over the ridge. What a sight! The telegraph station, the stage station, and the blacksmith shop were in flames. Hundreds of Indians milled around in the two-mile open stretch from the cliff to the military post.

Greatly outnumbered, with no place to retreat, the fifteen-man party had only one choice. They must make a run for it to the fort, right through the midst of the Indians. The smoke from the burning hay that hung near the ground offered partial concealment. Not expecting any soldiers to arrive, the Indians might think that the little band was the advance detachment of a large force.

Before dashing out from the protection of the cliff, the soldiers loaded the howitzer with what they called "canned trouble." This was canister, a can filled with small iron balls that was rammed

Indians attack Butterfield's Overland Dispatch Coach as it dashes for safety with guard and passengers firing at the attackers.

down the gun barrel. When the gun was touched off, the iron balls flew over a wide area like oversize buckshot.

"Ready!" shouted Captain O'Brien. "Charge!"

With sabers drawn, the party spurred their horses forward at a fast run. The men on the stagecoach laid the whip to their team, shouting and firing their rifles at the Indians. Past the burning stage station they dashed, exchanging shots with the Indians. For a moment the caravan hesitated while a soldier drove home into the howitzer the piece of telegraph wire that served as a priming wire. A barrage of canister zinged through the swirling Indians. At almost the same moment the soldiers in the fort, who had been watching the action, fired their howitzer toward the oncoming party. The Indians in between found themselves under fire from both directions. Bewildered by this unexpected turn of events, they spread out in disordered confusion.

Through the center of the Indians the party galloped in a swirl of dust and smoke. As they neared the sod walls of the fort, soldiers inside threw open the gate and the party dashed in to safety. Not a man had been hurt. Fifteen men had pulled a perilous bluff through hundreds of Indians, and got away with it.

Everyone in the area of Julesburg was safe. But the telegraph line was smashed. Getting it back into operation, with the Indians still on the warpath, was a dangerous assignment.

CHAPTER 8

The Wire Talks Again

That night the bitter February cold on the open Plains stung the faces of white men and red men like a thousand needles. Attack by the Indians on the garrison appeared imminent to the men barricaded in the small military post. The soldiers set up hourly watches, while Captain O'Brien and Lieutenant Ware climbed into the top of the haystack that rose above the wall as lookouts. They burrowed holes in the straw for protection against the cold, and from their half-concealed positions kept a close watch for Indian war parties.

Across the river they saw an awesome sight. A few miles west of Julesburg the Indians had attacked a wagon train; among the loot was a wagon full of whiskey kegs intended for Denver. The warriors had driven this wagon load to Julesburg with captured oxen, pulled it across the sanded path on the river ice to the north bank of the Platte, and opened the liquor. Soon they were in the midst of a drunken orgy around a huge fire. They danced scalp dances for the gruesome trophies they had taken in recent raids, sang, and beat their drums.

For fuel, the Indians chopped down telegraph poles and dragged them with their horses to the fire—the very poles that Creighton's crew had put into place with such high spirits the day they began construction at Julesburg in 1861. The revelry went on long toward dawn. Meanwhile small groups of drunken Indians skulked around the walls of the fort in the darkness, looking for a way to get in. One shot a flaming arrow into the haystack in which O'Brien and

Ware had concealed themselves. Flames flickered from the hay and it appeared that the stack would burst into a pyramid of fire and smoke. Seeing the peril, a soldier dashed to the hay with a bucket of half frozen water and doused the fire as it gathered headway. The two officers stuck to their post through the night.

Dawn came without an assault. By the first light, the garrison saw the Indians withdrawing up Lodgepole Creek to the north. Before them they drove the captured livestock on which they had loaded stolen sacks of corn and flour, rich loot for their skimpy winter diet. Trails of white stretched behind the animals from torn flour sacks.

From Julesburg the cross-country telegraph ran east to Fort Kearney and Omaha, northwest to Fort Laramie and Salt Lake City. To the southwest ran the branch line to Denver. Poles and wire were destroyed for miles toward both Fort Laramie and Denver. Fortunately only a short stretch of wire had been pulled down toward the east. This was quickly repaired that morning. In the fires set by the Indians at the telegraph station, however, the settlement's vital telegraph instruments had been destroyed. Without them, the wires were useless.

Even this failed to daunt the operator assigned to Julesburg, a young man named Holcomb. He was determined to get out word of Julesburg's plight. While the Army officers watched in astonishment, he drove an ax into the ground, grasped the ends of the telegraph wire in his gloved hands and touched them against the iron part of the ax in a dot-and-dash pattern. In this laborious makeshift manner he succeeded in opening and closing the circuit in a way that Morse had never dreamed about. But then, Morse had never been caught on the frontier with hostile Indians lurking on the horizon. Such circumstances were great stimulators of ingenuity.

"Indians attacked Julesburg," the operator messaged. "Send help."

Amazingly, the message got through to the next station east. Holcomb put the wires into his mouth, and as the long and short electrical impulses of the return message came over the wire against his flesh, he deciphered them.

"Get ready to follow. Am coming. Livingston."

Although the Julesburg garrison hadn't been aware of it, a large Army detachment under Colonel Livingston was less than thirty miles away that morning. The Indians had learned about the presence of the colonel's force, which was why they had withdrawn so unexpectedly. Some hours later Livingston arrived after a forced march with four hundred cavalrymen, four artillery pieces, and, almost as important, a set of telegraph instruments.

Scouting parties of cavalrymen made forays from the fort and found that the Indians had withdrawn some distance. Colonel Livingston organized two repair parties. He decided that getting the telegraph line in operation again to the West Coast, and to Denver, was more urgent than sending his troops in pursuit of the Indians. Hundreds of government and private telegrams were piled up in San Francisco, awaiting transmission across the country. In the raw frontier city of Denver a hundred miles southwest of Julesburg, the settlers were cut off from the world, anxious and uncertain about where the Indians had gone. Nothing could be done to repair the lines, however, until a fresh supply of telegraph poles could be hauled in. The nearest stockpile was at Cottonwood Springs about a hundred miles east of Julesburg.

Over the wire crackled Colonel Livingston's orders to Cottonwood Springs: "Load wagons with the lightest poles you have and bring them to Julesburg as fast as the animals will go. Keep going night and day."

The first wagons in the train carrying the poles rolled into the Julesburg fort at ten o'clock at night two days later, lanterns glimmering in the midwinter blackness. Weary as they were, there was to be no rest for the drivers or the animals. "You are going out again right away," the men were told, an order that set them to swearing in disgust. But orders were orders. They went.

Escorted by a squad of nearly fifty cavalrymen who hauled along two mountain howitzers, one wagon train headed out along the wrecked telegraph line toward Denver. Squads of four men were organized. Two men had picks, one a shovel, and the fourth held the horses. The squad found the burned stump of a telegraph pole in the darkness. Two men drove their picks into the pole to shake

The soldiers who repaired the telegraph lines after the Indians attacked Julesburg in February, 1865, weren't lucky enough to have a post hole digger, used later when the telephone poles were installed. They had to dig the poles in by hand, under constant threat of attack.

it, and the third pried out the stump with his shovel. The men heaved a fresh pole from the wagon, dropped its end into the vacant hole, and tamped it down.

"Mount! Forward! Gallop! March!" snapped the order from the squad leader.

The four men rode ahead to another pole. Similar groups of four had been sent forward until a soldierly game of leap frog was in progress. After each squad set a pole, it galloped ahead to the next vacant burned-out one. The moon had risen and shed a cold dim light on the strange procession. Behind the pole wagons came a light wagon carrying telegraph wire, insulators, and instruments. At each newly set pole, a soldier climbed up and fastened the wire to the insulator on the crossarm. From time to time, in order to take up the slack in the wire, the end of it was attached to the wagon and pulled taut by the wagon team of mules. The stumps that were pulled up were tossed into the pole wagons for firewood.

On the work went, hour after hour. At dawn the soldiers took a break for breakfast, then resumed their task, which continued the entire day. Hanging over the operation was the fear of an Indian attack; clusters of Indians on horseback were visible on the hills near the horizon, watching. While mounted Army scouts circled the work train, artillerymen stood by their howitzers in the center of the procession, on the alert. From time to time the telegraph operator accompanying the expedition attached his instruments to the end of the wire and tapped out a report on the progress and the activities of the watchful Indians to headquarters in Julesburg.

Twenty-five miles from Julesburg the Denver-bound party found that the major destruction of the line by the Indians had ceased, but for miles beyond that point occasional poles and stretches of wire were missing. Finally, at Valley Station, fifty-two miles southwest of Julesburg, the operator established contact with the telegraph station in Denver. News that the telegraph was working again and that wagon trains soon would get through caused joy and relief in that city.

In a period of forty hours, the soldiers rebuilt eight miles of the telegraph line, strung up another twenty-one miles of damaged wire, and marched fifty-two miles under threat of an Indian attack.

Old photograph shows workmen erecting telegraph poles, providing communication between the railhead of the Union Pacific and its headquarters in Omaha.

Similarly, soldier work crews advancing northwest from Julesburg up Lodgepole Creek found the line to Fort Laramie and Salt Lake City destroyed. It was the same routine—swing the picks, dig out the burned pole, plant the new one, attach the wire, then "Mount! Forward!" until the line was restored. This wasn't the kind of war their companions in the Union Army were fighting back East in the closing months of the Civil War, nor did it win any medals, but the men involved in it showed a special kind of valor.

Once again the overland telegraph was working. News of the Union victories and of General Robert E. Lee's surrender to General Ulysses S. Grant at Appomattox Court House flashed along the wire, causing jubilation in the faraway cities of California. The dreams and ambitions of the men who built the telegraph to the West were not fulfilled, however. Still greater spaces awaited conquest by the lightning wire.

CHAPTER 9

On to Russia

When President Abraham Lincoln died from an assassin's bullet in Washington on the morning of April 15, 1865, after being shot the previous evening at Ford's Theater, the tragic news was received in San Francisco within an hour over the transcontinental telegraph. Word of the assassination did not reach London and the European capitals, however, until the steamship *Nova Scotian* docked in England on April 26, eleven days later. While the three thousand miles from the national capital to the Pacific Coast were spanned almost like lightning, the message traveling roughly the same distance across the Atlantic Ocean could move only as fast as the ship that carried it.

One of man's ambitions in the middle of the nineteenth century was to lay a submarine cable across the Atlantic through which words could be telegraphed along the ocean bottom from one shore to the other. Many men talked about that dream. As so often is the case, however, the dream became such an obsession to one man in particular that he committed his life and fortune to making it come true.

His name was Cyrus W. Field. He knew how it was to be hailed as a hero, ridiculed when things went wrong, and ultimately praised again when success finally came. Field's up-and-down struggle to lay the Atlantic cable led to a strange and almost forgotten adventure halfway around the world by rival men who gambled three million dollars on a fantastic telegraph scheme, and lost.

After two failures, Field and his organization successfully laid

Cyrus W. Field

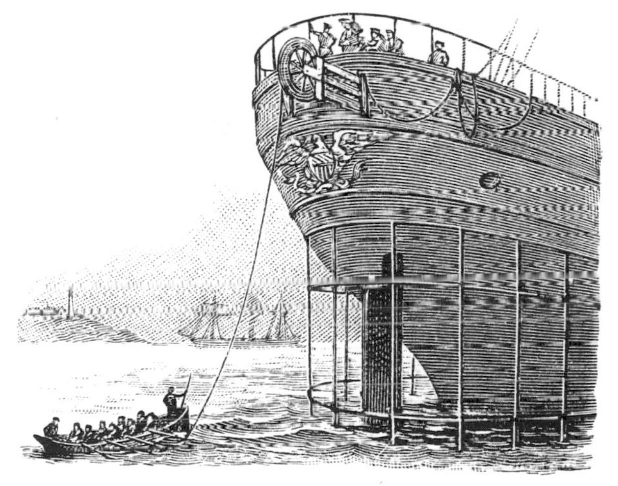

Laying the Atlantic Cable

a cable across the Atlantic during July and August, 1858. Great excitement greeted the news. Americans were jubilant that at last they would have quick communication with Europe, from which they drew so much of their culture and history. But the joy was short lived. The cable didn't work properly. Only a few garbled messages were transmitted because the underwater wire was improperly insulated and the ocean waters ruined the electrical circuit. The people who had cheered Field so lustily turned against him.

During the Civil War, the resources of the United States were far too heavily involved in the fighting for anyone to make a fresh

In this old cartoon, Neptune says to the mermaids, "Ahoy, there! Get off that there cable, can't yer? That's the way t'other one was broken."

attempt at a scheme that probably wouldn't work, anyway.

About the middle of the war years, an imaginative man named Perry McD. Collins went to the high officials of Western Union with an elaborate proposal.

"The underwater cable hasn't worked and probably never will," he said. "Instead, I suggest that you build a telegraph system the other way around the world to Europe. You already have the transcontinental line working to California. Build the line north from California through western Canada and Alaska, across the Bering Strait to Siberia and connect it with the line the Russians are building eastward."

What Collins proposed was a line of galvanized iron telegraph wire sixteen thousand miles long that would carry a message from New York two-thirds of the way around the world to reach Paris and London. The distance was more than five times as great as that required for a cable under the Atlantic Ocean. Such are the oddities of geography, however, that in the entire sixteen thousand miles the telegraph wire would have no oceans to cross. The widest water barrier was the relatively narrow Bering Strait between Alaska and Siberia. At that time, Alaska still belonged to Czarist Russia. The Russian government was friendly to the United States.

As the persuasive Collins explained his plan to the financiers who ran Western Union, those hard-headed men caught his enthusiasm. Their original objections—that the distances were huge, that the telegraph wire must be strung across thousands of miles of virtually unexplored wilderness under conditions of extreme hardship—melted away. As they ran their fingers across maps spread on a table in their New York office, the plan seemed reasonable. Hadn't they already put the line through to San Francisco when skeptics said it couldn't be done? Visions of controlling all the messages flowing between the United States and Europe, and the rich profits this business would bring them, turned their original doubts into agreement.

"What if Field succeeds in laying his cable?" one asked. "That would destroy our project."

"He'll never do it," Collins assured them. His optimism was

A plan to circle the world with wire was born during the Civil War in the mind of Perry McD. Collins.

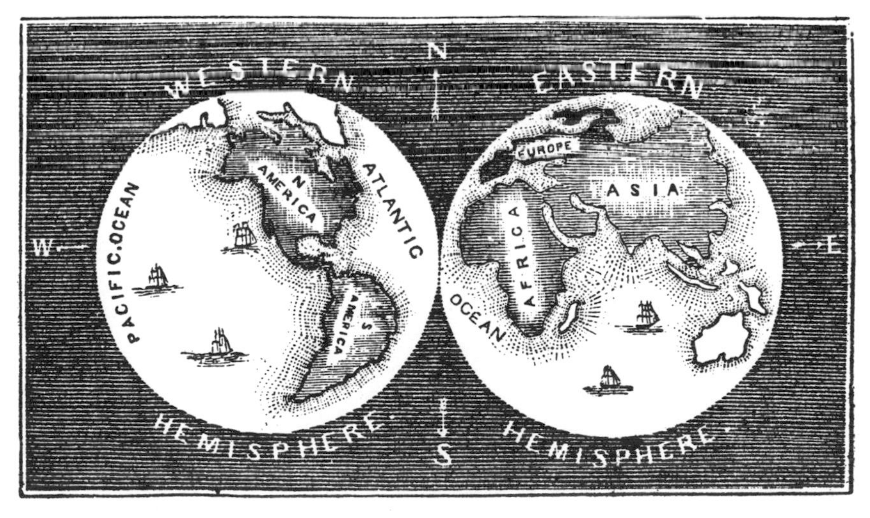

reinforced in the summer of 1865, shortly after the war ended, when Field made another effort to lay the cable. This time the line broke in mid-Atlantic. Its ends were lost under hundreds of feet of water.

Far away in California, the bizarre adventures of the Russian-American Telegraph Company were starting. Parties of men set out by land and by sea to string a wire through the back door of the world.

From San Francisco, work crews built the telegraph line north up the Pacific Coast through the redwood forests and across the streams and valleys of Northern California, past the looming white-tipped peak of Mount Shasta, and into Oregon. Here there was ample timber for poles. Indeed, the problem, the opposite from that faced on the Great Plains, was running the wire through areas of dense forest. From Oregon the parade of poles stretched slowly up through the State of Washington until it reached New Westminster, just south of the city of Vancouver. From there the line would run up the Fraser River Valley. A side branch was built across the channel to the southern tip of Vancouver Island to Victoria, a picturesque city that had been started as a Hudson's Bay Company fort and trading post twenty-five years before the telegraph line approached it.

Here a new obstacle developed. Getting the line across the strait to Victoria required sinking it underwater onto the bottom of this narrow arm of the Pacific Ocean. The builders encased the wire in a new waterproof substance, gutta percha, the material from which golf ball covers were made later. A shipload of cable for the underwater project was sent from the East Coast on the long trip around South America but was lost at sea in the turbulent waters off Cape Horn. Orders for replacement cable were telegraphed from San Francisco, but months of delay resulted before the new shipment reached Canada. Once the cable arrived, it was laid across the strait successfully. A telegraph sounder hooked up in Victoria received the first messages from the United States.

Command of the entire project was given to Colonel Charles S. Bulkley, who set up headquarters in San Francisco. He had been

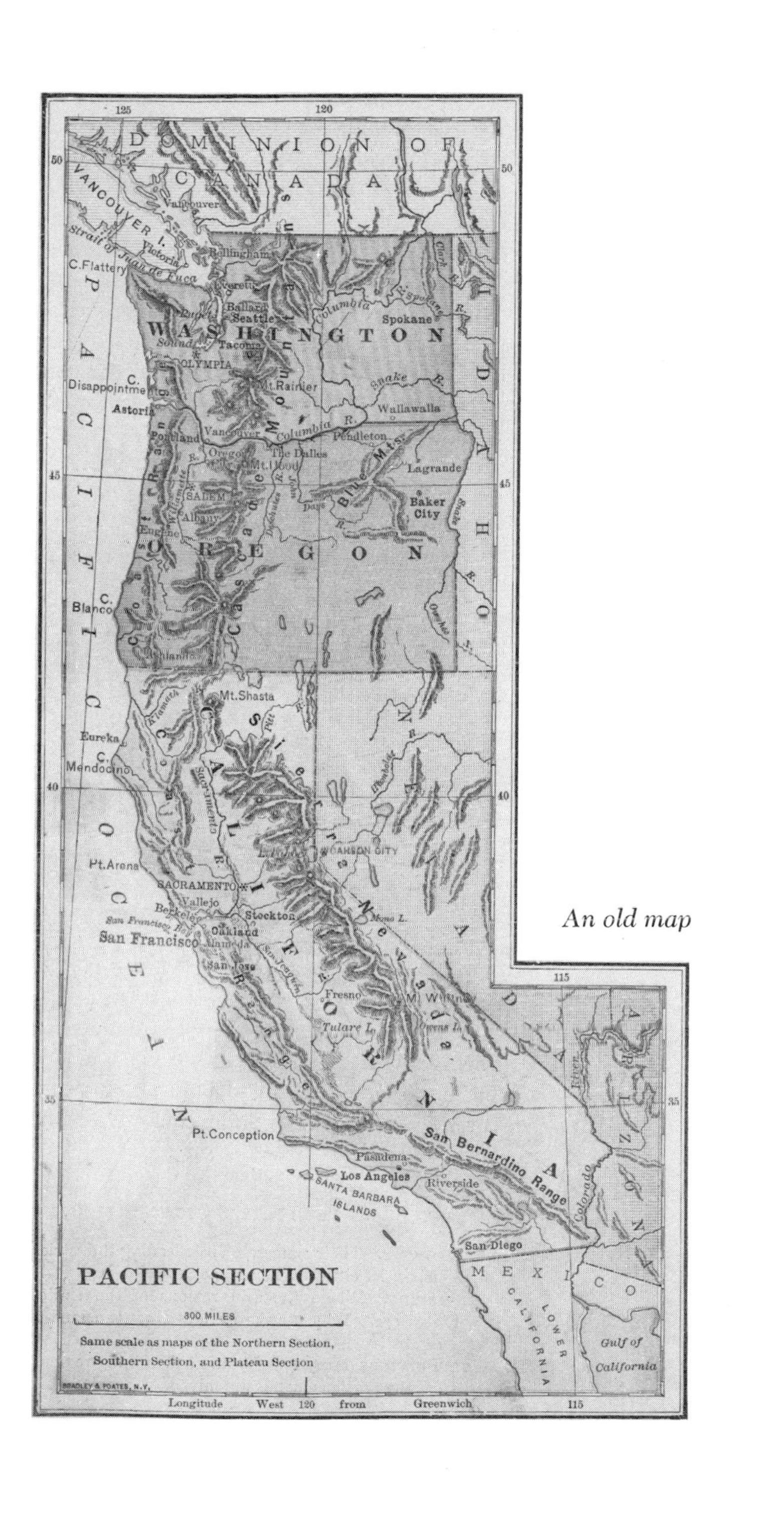

PACIFIC SECTION
300 MILES
Same scale as maps of the Northern Section, Southern Section, and Plateau Section
Longitude West 120 from Greenwich
DOMINION OF CANADA
WASHINGTON
OREGON
Vancouver
Vancouver I.
Victoria
Strait of Juan de Fuca
C. Flattery
Bellingham
Everett
Ballard
Seattle
Spokane
Tacoma
OLYMPIA
Mt. Rainier
Wallawalla
Astoria
Portland
Pendleton
The Dalles
Lagrande
Baker City
SALEM
Albany
Eugene
Blue Mts.
C. Blanco
Mt. Shasta
Eureka
C. Mendocino
CARSON CITY
Pt. Arena
SACRAMENTO
Vallejo
Berkeley
Oakland
Stockton
San Francisco
San Jose
Fresno
Mt. Whitney
Tulare L.
Mono L.
Pt. Conception
Pasadena
Los Angeles
Riverside
SANTA BARBARA ISLANDS
San Bernardino Range
San Diego
MEXICO
LOWER CALIFORNIA
Gulf of California

An old map

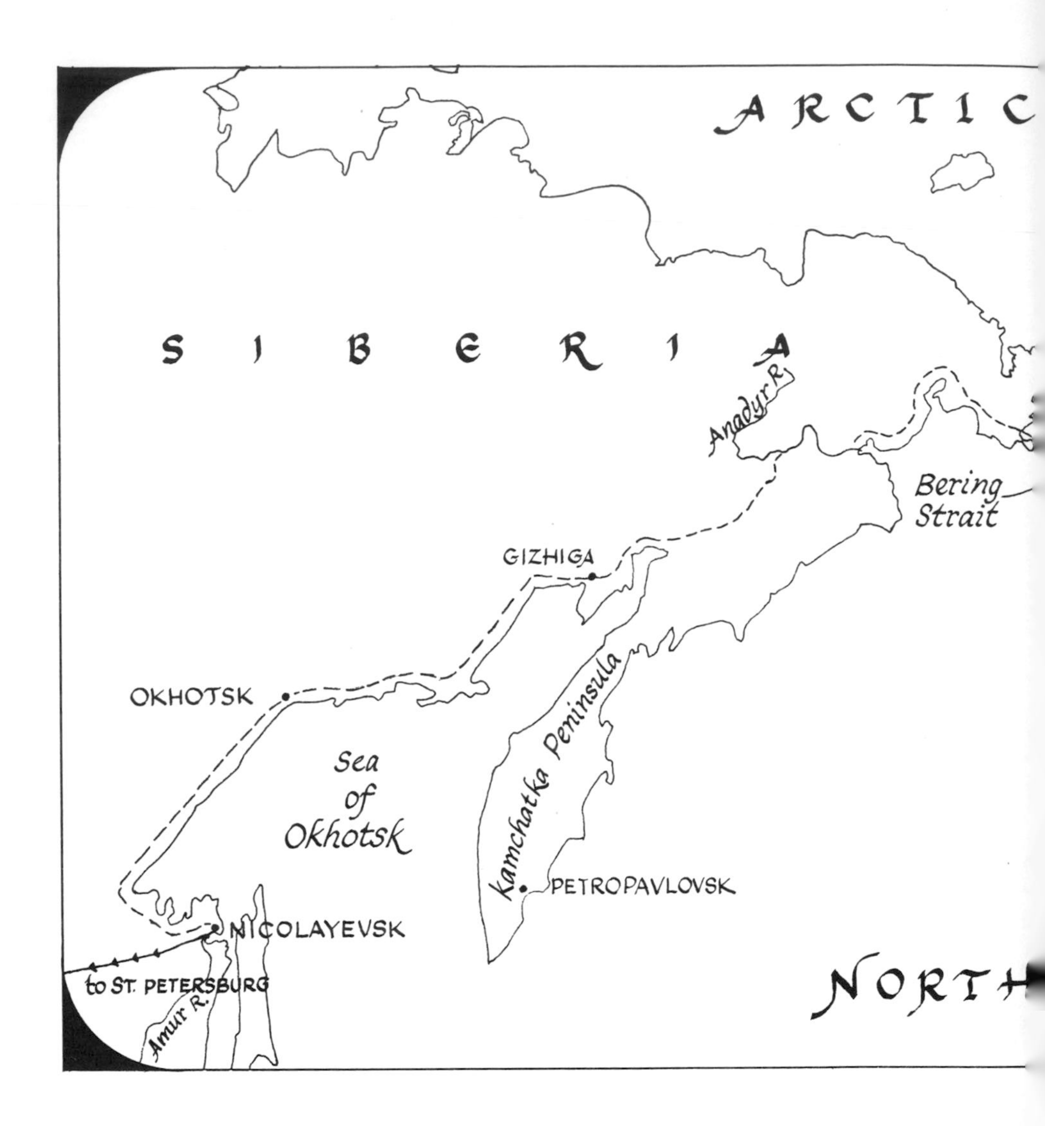

superintendent of a military telegraph section in the Union Army. After studying the best available information about the immense, wild territory involved, he established operations in three divisions. One force based in Canada was ordered to build the line twelve hundred miles up the Fraser River in British Columbia and on to the edge of Alaska. A second work gang was assigned the task of constructing the line nine hundred miles across the little-known

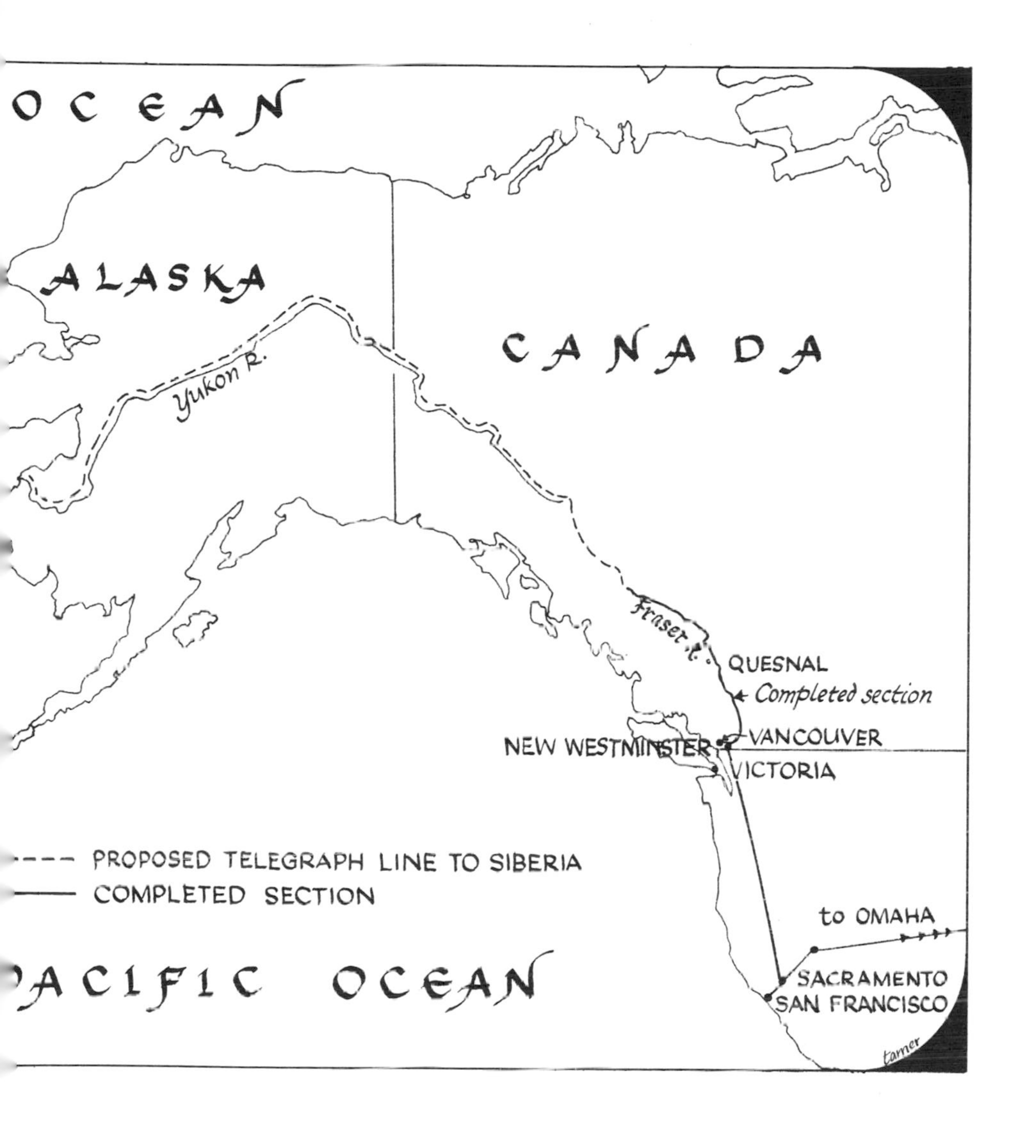

mountains of Russian Alaska to the Bering Strait, where a ship would lay a cable across to the Siberian coast. The third force had the most dangerous task of all—to land in northern Siberia and build a telegraph line from the Bering Strait eighteen hundred miles southwest across the Arctic tundra near the shore of the Sea of Okhotsk to the mouth of the Amur River, which flows into the sea from deep in Siberia. At this point the American line was to be hooked up with a line seven thousand miles long that the Russians were building from St. Petersburg (now Leningrad) across Siberia. From St. Petersburg, a telegraph line already ran to western Europe and Great Britain.

The immensity of the scheme was staggering. Only business leaders with tremendous optimism and inadequate knowledge of the harsh geography they sought to conquer would have dared to undertake it. They were asking their men to build a telegraph line through territory where the temperature dropped to 68° below zero, nights were more than twenty hours long in winter, travel was impossible in places during the summer and only by dogsled or reindeer in winter, and nobody lived except a handful of natives, mostly nomads. Many portions of the route had never been visited by western men.

No lack of volunteers was evident, however. Colonel Bulkley had hundreds of men in San Francisco from whom to choose his three forces. Mostly they were adventurous young fellows, some of them miners from the Mother Lode country of California, others discharged veterans of the Civil War who had drifted to the West to start a new life. He had the full cooperation of Great Britain, ruler of the Canadian territory through which the line would run, and of Russia, whose flag flew over both Alaska and Siberia. The construction work was to be done entirely by the American company.

In midsummer of 1865, a few weeks after the last shots of the Civil War had been fired in Virginia, Bulkley's three task forces set to work.

A thousand men began construction in British Columbia. They had two advantages, of sorts. A skimpy kind of trail had been opened through part of the Fraser River Valley wilderness by men who had gone through there during a gold rush in 1855. And second, as they built each additional few miles of line, an operator could hook up his instruments at the end of the wire and exchange messages with the outer world. Far away from civilization as these men were, they kept track of what was happening back home.

At places, the poles had to be sunk into holes blasted in the rock. Steep cliffs dropped down to the river's edge on both sides.

Old prints show use of dogs for transportation in the Far North.

The work pushed ahead, nevertheless, during the last months of 1865, to the remote settlement of Quesnel, more than four hundred miles up the Fraser River from Vancouver. During the winter when heavy snow closed in, survey teams explored the route farther north toward the Arctic Circle on sleds drawn by teams of dogs and reindeer.

Exploring parties also cut a right-of-way fifty feet wide across portions of Alaska and put up poles. Slowly, laboriously, the way for the telegraph was being opened to the Bering Strait, although hundreds of miles of wire remained to be strung.

Reindeer were used for transportation in the Arctic, too.

While the American crews in British Columbia and Alaska had an assignment that taxed their strength and persistence, the third party of the Russian-American Telegraph Company—the one sent to windswept Siberia—encountered conditions almost too extreme to endure. Indeed, before their adventure ended, one man had died and another had taken his own life in despair.

Thanks to George Kennan, a twenty-year-old telegrapher from Cincinnati, we have a fascinating account of the attempt to build the American telegraph line across eighteen hundred miles of Siberia. Kennan kept a diary, later published in book form, that contains stories of the party's experiences in the forbidding land.

On the bright afternoon of July 1, 1865, a party of three Americans, including Kennan, and a Russian army major sailed from San Francisco on a creaky Russian trading vessel, the *Olga.* Their destination was the Kamchatka Peninsula of Siberia and the Okhotsk Sea, an arm of water cut off from the far northern Pacific Ocean by the southward-pointing finger of the peninsula. Seven weeks later—weeks filled with seasickness, discomfort, and boredom—the little brig arrived at Kamchatka. The Russian, Major Abaza, and Kennan went ashore there, and the *Olga* sailed on across the Okhotsk Sea to deposit the other two men on the Siberian mainland. Split up, the four were to explore segments of the Siberian coastal region from the Bering Sea down to the link-up spot with the Russian telegraph at the mouth of the Amur. Supplies and workmen were to follow from San Francisco by ship the next spring. Once the route was explored, the advance parties were to hire Siberian natives to cut trees for telegraph poles. The plan sounded fine, on paper.

Kennan and Major Abaza rode north up the Kamchatka Peninsula on horseback, stopping in villages whose natives were amazed to see an American for the first time.

As they entered one village they were greeted with exceptional shouting and gesticulating. The head man of the village bowed and bowed to Kennan in such a worshipful manner that the young American was embarrassed.

"What is the matter? Is this settlement insane?" he asked. Suddenly the major, speaking to the head man in Russian, broke into a grin.

He told Kennan, "The natives took you for the Emperor."

As it turned out, a messenger had been sent ahead with a note from the Russian governor giving the names of the visitors. The note listed Kennan as "telegraphist and operator." Nobody in the settlement had any idea of what "telegraphist" meant, because they had never heard of a telegraph. And the head man misread "operator" as "imperator," meaning Emperor! He had rushed around the village, spreading word that the natives were about to be visited by the great emperor who lived eight thousand miles away in St. Petersburg. For a few moments the youthful American had a taste of what it meant to be royal, even if it was a case of mistaken identity.

That winter of 1865-66, Kennan's portion of the surveying party traveled to the northeastern corner of Siberia, almost to the Russian shore of the Bering Strait, to locate a route for the telegraph line. Living conditions were appalling. The temperature dropped to 68° below zero and blizzards whined across the frozen steppes. Only four hours of daylight occurred in each twenty-four while the rest was pitch blackness. The few native settlements were about two hundred miles apart.

Far away in Western Union headquarters in New York, the Russian-American telegraph plan looked good on the maps. For the men in the field doing the work, it was close to torture. They could not dig in at a native settlement and wait for spring, however, because their exploring had to be done during the frozen months. In summer, the melting snow exposed great moss tundras so soggy that they were impassable. Across great stretches of this tundra there were no trees. One of Kennan's assignments was to hunt for routes where sufficient clusters of gnarled Arctic trees survived to be cut for poles—a familiar problem!

At one settlement near a grove, Kennan hired the natives to chop trees. The poles were to be twenty-one feet long and five inches in diameter at the top. Three months later, as the first signs of spring developed, Kennan drove his nine-dog sled back to the settlement. He found five hundred poles cut and stacked, almost every one of them at least twelve inches in diameter at the top and so huge that a dozen men could hardly move one.

"Why didn't you cut them the size I said?" he asked.

A native replied, "We thought you planned to build a road on the tops of the poles. We knew that poles as thin as you wanted would not hold up a road." Such was the native concept of the telegraph line they were supposed to help build!

When caught at night on the steppes, the explorers survived by using the lore developed by the childlike but clever Siberians. Three dog sledges were drawn up like three sides of a square about ten feet across. The dogs curled in balls in the snow, their breath making small clouds of steam. The sledge drivers shoveled the snow out of the square and spread twigs on the ground. Next they put down shaggy bearskins, on which the men placed their fur sleeping bags. Masses of a tundra vine were pulled up from beneath the snow, piled high at the open end of the square, and set afire. Encased in furs and warmed by the tea they had brewed, the men settled into their little snow bivouac for a few hours of sleep, if not snug at least sufficiently protected to come through alive. In the morning when they awakened their eyelids frequently were frozen shut.

Landing the Atlantic Cable at Heart's Content

By spring of 1866 the Americans had explored almost every bit of the route. They had engaged natives to cut more poles and had constructed huts to serve as telegraph relay stations. They gathered in a village on the Sea of Okhotsk to await the arrival from San Francisco of the promised supply ship with the rest of the equipment to build the great Siberian telegraph.

Weeks passed, but no ship arrived. Totally out of touch with home for a year, the little party could do nothing but wait idly and wonder what had gone wrong.

Not until late August did two supply ships anchor off shore, two months en route from San Francisco. When the seamen opened the holds, all the supplies for which the advance party had waited so eagerly were there. Also aboard were sixty Americans sent from San Francisco as construction crews.

"At last we're ready to build!" Kennan's party rejoiced.

Had they known the truth, their happiness would have withered into regret. Just a month earlier, on the other side of the world, Cyrus W. Field had succeeded in laying the Atlantic cable. Messages were clacking steadily back and forth beneath the ocean from

Europe to New York. No longer was there any need for the Russian-American telegraph, but the men in Siberia had no way of knowing this. Eagerly they plunged ahead in a cause nobody was able to tell them had already been lost.

CHAPTER 10

"The Craziest Thing Ever"

When news that the Atlantic cable was working clicked up to the head of the telegraph wire in the wilds of British Columbia, the hundreds of workmen there realized immediately that their jobs were finished. Within a few days they abandoned camp and headed south toward civilization. Behind them they left abandoned tools and supplies—and a working telegraph line up the Fraser River Valley that reached some five hundred miles toward Alaska. A line of waiting poles reached still farther.

Western Union kept the line in operation to Quesnel, the one thin strand of quick communication to that remote settlement. In 1870 the company sold the telegraph line to the Canadian government. Beyond Quesnel the wire and poles were abandoned but stayed in place until weather and the local Indians took their toll. The impoverished Indians found the wire too inviting to be resisted, especially since nobody was around to stop them. They cut it from the poles. Soon it found uses its manufacturers had never imagined, as nails to build shacks, as points for fishing spears, and for traps to catch wild animals. Even more ingeniously, the Indians used lengths of the wire as thongs to bind together the logs in crude semi-suspension bridges they built across the rapid streams. For many years during the latter part of the nineteenth century these telegraph-wire bridges bore the weight of passing Indian parties.

Thirty years after the telegraph line to Russia was abandoned in Canada, the Klondike gold rush developed in the late 1890's. Again, thousands of men eager for riches plunged into a rugged, lonely region. Many reached the gold fields by going up the Fraser River Valley, far beyond Quesnel. Waiting for them was the right-of-way cleared three decades earlier by the Russian-American Telegraph Company. At last it had a use; the Canadian government strung a wire along it to the mining camps of the Yukon. As late as 1905, some forty years after the original work was done, a survey party exploring an uninhabited area found a section of the original poles still standing, with rolls of telegraph wire nearby waiting to be put up.

Although the dream of a telegraph line to Russia was shattered, it had an unexpected benefit. Publicity about the project increased interest among American government officials concerning that little-known part of the continent. Secretary of State William H. Seward, in particular, became intrigued by Russian America, as Alaska then was called. Seward was a booster of the telegraph project, too. Reports of the telegraph builders' activities enlarged Americans' knowledge of this distant region and proved that construction work was possible there, despite atrocious weather several months of the year. With brilliant foresight, Seward realized that Alaska might some day be of immense value to the United States. So in 1867, the year after telegraph construction work was abandoned, he arranged for the United States to purchase the territory from the Russian Czar for $7,200,000. People hooted at the deal and called it "Seward's Folly." Imagine paying good money for that inhospitable wasteland! Today, as our forty-ninth state, Alaska has proven a thousand times over how wise Seward was.

All of this was unknown to the American telegraph builders across the Pacific in Siberia in 1866. They had no idea of what was going on back in the United States. All they knew was that their orders were to build the telegraph line. They had their supplies, at last. On with the job!

Another supply ship arrived in late September, ran aground in a storm, and was unloaded just in time before the zero weather filled the Sea of Okhotsk with ice. Major Abaza had gone to Yakutsk, the nearest city, to hire six hundred Siberian laborers and buy three hundred horses to pull the poles that were being cut. Optimism ran high. The men composed a lusty song to sing as they cut poles:

In eighteen hundred and sixty eight
  Hurrah! Hurrah!
In eighteen hundred and sixty eight
  Hurrah! Hurrah!
The cable will be in a miserable state,
  And we'll feel gay
When they use it to fish for whales.

Little did that isolated crew know that they, not the cable, were the ones in the miserable state.

Through that second winter, of 1866-67, the Americans, with the help of natives, cut poles until they had nearly twenty thousand of them stacked up, ready to be dragged into position. They worked on snowshoes, in temperatures that fell as low as 60° below zero in the brief midday hours of daylight, so swaddled in furs that swinging an ax was difficult. One group of Americans, based in

Siberian sledges and a Siberian village in winter

the far corner of Siberia near the Bering Strait, almost starved when their supplies ran low. They were saved when nomad tribesmen gave them part of their own small store of reindeer meat.

As winter turned into the sudden Siberian spring of 1867, the work moved steadily ahead. The men talked about having the entire line to St. Petersburg in working order by the end of 1869.

On the evening of May 31, a whaling ship appeared off the coastal settlement where Kennan was staying. The next morning he and others rowed to the anchored vessel and found that she was the *Sea Breeze* of New Bedford, Massachusetts, commanded by a Captain Hamilton. The captain was amazed to discover that the fur-clad men were Americans.

"What are you doing up in this God-forsaken country?" he exclaimed. "Have you been shipwrecked?"

"No, we're up here to build a telegraph line."

The captain snorted in amazement, "A telegraph line! Well, if that ain't the craziest thing I ever heard of! Who's going to telegraph from here?"

Kennan explained about the line they were building from the United States to Russia. "What news is there of the Atlantic cable?" he asked.

The captain smiled, apparently believing that he was giving them good news. "Oh, yes, the cable is laid all right. Works like a snatch-tackle. The 'Frisco papers are publishing every morning the London news of the day before."

This was staggering information. Their work and suffering had been in vain. At first it was difficult for the men to realize how utterly useless their project had become. All that effort wasted! Disheartened, Kennan and his companions started ashore. When Captain Hamilton recognized what bad news he had brought the men, he tried to cheer them as best he could. He gave them a supply of fresh vegetables and fruit—the first they had tasted in nearly two years—and the San Francisco newspapers he had on board. Once ashore, the Americans roasted the potatoes they had been given, sucked on oranges, and read the newspapers. These dated back many months, but the news was fresh and exciting to the long-isolated men. In a copy of the San Francisco *Bulletin* of

George Kennan in Siberia

October 15, 1866, more than seven months old, one of them found a brief telegraphed dispatch from New York: "In consequence of the success of the Atlantic cable, all work on the Russian-American telegraph line has been stopped and the enterprise has been abandoned."

Little wonder Kennan and his friends were downhearted. Still, they had no official notice from their company and no transportation home, either. There was nothing to do but wait and wonder.

Six weeks later another small vessel arrived in the Sea of Okhotsk carrying formal word from Western Union officials in New York to the Siberian party to sell their supplies as best they could and make the long trip home.

The men took stock of what they had: twenty thousand telegraph poles and crossarms brackets, miles of wire, thousands upon thousands of glass insulators to hold the wires to the poles, building

Patrons of the American Telegraph Company in New York prepare messages for dispatch to Europe on the newly completed Atlantic Cable.

tools, and food supplies. How in the world could they sell these things to the few thousand Siberian natives who didn't need them and had few enough rubles to buy anything, living as they did largely by barter?

The Americans found themselves cast in the role of salesmen instead of builders. Cleverly, they used a traditional sales technique by creating artificial wants, convincing the natives that they needed things they didn't. Kennan recalled later, "We sold glass insulators by the hundred as patent American teacups, and brackets by the thousand as prepared American kindling wood. We offered soap and candles as premiums to anybody who would buy our salt pork and dried apples, and taught the natives how to make cooling drinks and hot biscuits in order to create a demand for our redundant lime juice and baking powder." Those natives who bought shovels were given frozen cucumber pickles as a bonus. Perhaps the man who later supposedly tried to sell refrigerators to Eskimos got his idea from this piece of super-salesmanship!

What became of those twenty thousand telegraph poles stacked on the frozen tundra never was disclosed. Probably they eventually were chopped for firewood to burn in the native huts.

Thus ended a grandiose dream that became a frostbitten nightmare. We are left wondering, would the Arctic telegraph really have worked if Field had failed again with his Atlantic cable and the line had been completed to St. Petersburg? That is one of history's tantalizing unanswered questions.

CHAPTER 11

Here Come the Trains

Far out on the sagebrush plain of Utah north of the Great Salt Lake at Promontory Point, Governor Leland Stanford of California in frock coat and stiff hat lifted a silver-plated maul. Proudly he stepped toward a loose piece of rail in the single track that spanned the huge empty area. The midday sun of May 10, 1869, was warm and springlike, a good omen for an historic moment in American history, completion of the first transcontinental railroad.

A short time earlier two railroad engines had steamed slowly toward each other until their cowcatchers were only a few yards apart. Facing east was the Central Pacific's Jupiter, a woodburner with smokestack round like a black iron balloon, its shiny brass trimmings accentuated by a coat of red and yellow paint. Pointed west toward the Pacific Ocean was the Union Pacific's No. 119, one of the first coal-burning locomotives west of the Mississippi River. Soldiers lined the area and a military band played march music. Dozens of the men who had built the railroad during years of struggle clung to the tops of the locomotives to get a better view.

As one of the men who headed the Central Pacific, Stanford had been given the honor of driving the final spike of California gold to complete the railroad.

It was a moment long and eagerly anticipated by men and women on both the Atlantic and Pacific Coasts. Yet the site was in such a remote section of the country that only a few persons could witness the ceremony. The word that the final spike had been driven would

Two photographs of ceremony marking completion of the first transcontinental railroad, May 10, 1869. Locomotives from the East and the West meet nose to nose as men climb telegraph pole for a better view. News that the final spike had been driven was flashed to the world by the telegraph instrument on the table at right center of bottom photograph.

be spread from coast to coast instantaneously by the telegraph, however. It would be some of the grandest news ever transmitted along the galvanized iron strand of wire.

Alongside the track a table had been placed, and on it the Western Union operator from Ogden, Utah, W. N. Shilling, had put his sending key and receiving instrument. From the table ran wires up to the line on the telegraph poles that stretched to the horizon along the track, both to east and west.

Elaborate electrical arrangements had been made in many American cities to touch off celebrations the moment word arrived on the wire from Shilling that Stanford had driven the final spike. A magnetic bell had been rigged to the receiving apparatus in the Western Union office at Fourteenth Street and Pennsylvania Avenue in downtown Washington. This was to start ringing when the electrical impulse from Shilling's sending key 2,400 miles away reached Washington. Similarly, a magnetic ball had been placed on a pole atop the Capitol. The electric impulse was to make the ball drop, so that members of Congress and others on Capitol Hill would know the news instantly.

The ceremony at Promontory Point began with a prayer. Knowing that waiting crowds all over the country were growing restless, Shilling tapped out, "We have got done praying. The spike is about to be presented."

Moments later: "All ready now; the spike will be driven. The signal will be three dots for the commencement of the blows."

Stanford raised the maul above his head and swung. As the tool arched down, Shilling tapped out a dot, starting the national celebration. An embarrassing moment followed. The governor had missed. Instead of hitting the golden spike, Stanford's maul smashed down on the rail itself. But there was no turning back. Shilling had completed his three-dot message and the celebration was on. When the red-faced governor did succeed in driving the spike into the specially made crosstie of California laurel a few moments later, Shilling flashed the word, "Done."

Cannon in New York fired a hundred-gun salute and the chimes of Trinity Church rang out the news. High officials of the federal government gathered in the telegraph room of the War Department

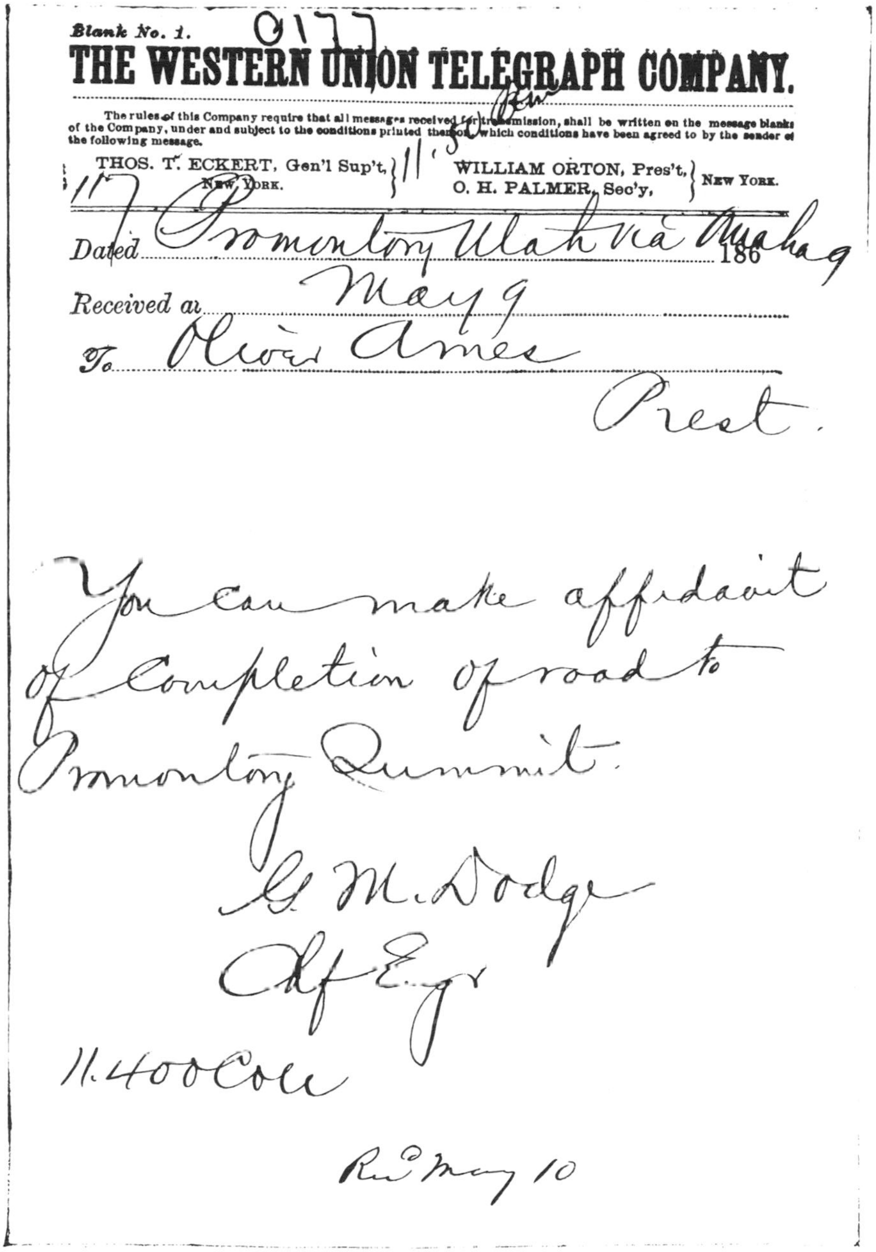

Blank No. 1. 0177

THE WESTERN UNION TELEGRAPH COMPANY.

The rules of this Company require that all messages received for transmission, shall be written on the message blanks of the Company, under and subject to the conditions printed thereon, which conditions have been agreed to by the sender of the following message.

THOS. T. ECKERT, Gen'l Sup't, NEW YORK. 11 30 Bw

WILLIAM ORTON, Pres't, O. H. PALMER, Sec'y, NEW YORK.

117 P

Dated Promontory Utah via Omaha 9 186

Received at May 9

To Oliver Ames

Prest.

You can make affidavit of Completion of road to Promontory Summit.

G M Dodge

Chf Engr

11.400 Coll

Recd May 10

Telegram announcing completion of the Union Pacific Railroad at Promontory, Utah, from Grenville M. Dodge, chief engineer of the Union Pacific Railroad, to Oliver Ames, president to get the word, and up on Capitol Hill after the magnetic ball dropped there was cheering. In Chicago an impromptu parade formed, seven miles long. Outside Independence Hall in Philadelphia a line of steam fire engines sounded their whistles while the Liberty Bell in the tower rang joyously. Perhaps the greatest happiness of all was in San Francisco, at the western end of the railroad. For the city at the Golden Gate, completion of the railroad meant the end of isolation from the East. A wire had been strung under the streets of San Francisco from the telegraph office out to Fort Point; there it was attached to a fifteen-inch gun. As the first dot of Shilling's message was received, the Fort Point gun fired the first of 220 rounds while simultaneously all the fire bells in the city sounded.

Old prints. Left: The transcontinental railroad track heads into the lofty Sierra Nevadas, with the telegraph line strung alongside it. Right: *Early train and the telegraph in Animas Canyon, Colorado.*

While the celebrations were for the completion of the railroad, the performance of the telegraph in carrying the news instantaneously from the remote desert of Utah to the entire country was noteworthy, too. Less than eight years had passed since Gamble's construction crews had strung the first cross-country wire a few miles to the south of Promontory Point. In 1861 the Pony Express was the fastest way available to carry a message across the great gap in the country. Now in 1869, the West had both a telegraph and a railroad.

Those poles along the railroad tracks signified something more, the fact that the telegraph had become an essential part of making the nation's trains run. Whenever a long distance railroad was built, the telegraph line accompanied it. In fact, most intercity telegraph wires followed the railroads and came to be thought of almost as part of the railroad by the general public. It had not always been that way.

In the early days of American railroading, and even for several years after the invention of the telegraph, the methods used to prevent collisions between trains coming toward each other on a single track were primitive. The usual way was to designate the train going one direction as the "superior" one, and a train in the other direction the "inferior." The superior train had a one-hour grace period to pass a designated point. If the superior train had not yet arrived after that hour, the waiting one sent a signalman ahead with a red flag. He walked for twenty minutes, then his train caught up with him, took him aboard and sent another man walking ahead, until he met the oncoming train. This prevented head-on collisions but often led to arguments, even fistfights, between crews when their engines came head-to-head as to which one should back up to the nearest siding.

Superintendent Charles Minot of the Erie Railroad was the man who discovered that trains could be dispatched and controlled by telegraph messages. A westbound train on which he was riding in New York State had been held up at a station named Turner's because the "superior" eastbound train had not yet passed. There was no way to know where it was. After the required hour of waiting had passed, an idea suddenly came to Minot. Instead of sending a brakeman ahead on foot and having the westbound train creep along behind him, Minot turned to the telegraph operator at Turner's station.

"Ask the operator at Goshen if the eastbound train has left there yet." Back came the reply; the train, running very late, hadn't even reached Goshen.

Whereupon Minot sent a telegram to the Goshen station master: "Hold the train for further orders." Turning to the conductor of his own waiting train, he wrote an order, "Run to Goshen regardless of opposing train."

This shocked the train crew. They had visions of a head-on collision. When the engineer refused to proceed, Minot climbed into the engine cab while the frightened engineer took a seat at the back of the rear coach.

No crash happened, of course. Indeed, when Minot brought his train to Goshen, the eastbound one still hadn't arrived, so Minot telegraphed "hold" orders ahead to Middleton and subsequently to Port Jervis, where the two trains arrived almost simultaneously. More than an hour in running time had been saved by the westbound train through this use of telegraph messages. A new era of railroading had begun. By the time the transcontinental railroad was completed, movement of trains by telegraphic orders was standard procedure. The telegraph operator was a part of the staff in virtually every railroad station, handling both railroad orders and private telegraph messages for Western Union. In small stations, the station agent doubled as telegraph operator.

One sleet-filled night around the turn of the twentieth century, the Overland Limited was steaming at high speed along the Union Pacific tracks in sparsely settled country forty miles west of Cheyenne, Wyoming, when a rail broke loose and jammed up through the bottom of the baggage car. This type of train accident

The railroad station at Cheyenne, Wyoming, in the late 1800's, showing the telegraph office

was not uncommon in those days of thin, poorly laid rails. The Overland Limited jumped the track, and its locomotive and nine cars piled together in a mass of smoking wreckage. It was a bitter black night, the train was miles from a station; no way was apparent to get out word of the wreck to the world. The next train wasn't due to pass for five hours. On the ground and in the splintered coaches were approximately 150 dead and injured. Medical help was needed, urgently.

Out of the wreckage of the baggage car crawled a man named Frank Shaley. A telegraph lineman, he was being sent up the line to look for reported wire trouble and had caught a ride in the baggage car. He was bleeding and badly injured, barely able to pull himself along the ground.

"Cut in on the telegraph wire," he urged the uninjured men who gathered around him.

By then, the transcontinental telegraph had grown from a single strand to many wires, so great was the business. None of the men knew which one to cut, or how to do it. Holding his bag of telegraph instruments and tools, Shaley muttered, "Lift me up there."

A rope was thrown over the crossarm of a telegraph pole, and a sling for Shaley was tied to it. In this manner he was hoisted up to the level of the wires. Faint and bleeding profusely, he cut the proper wire, grounded it, and attached his telegraph key. Helping hands from below propped him into something resembling an upright position.

Shaley tapped the key, trying to raise the operator at Cheyenne. At that late hour on a stormy night, not expecting any messages or trains, the operator had either stepped away from his desk or dozed off. For ten minutes the critically injured Shaley, half frozen in the sleet, pounded his key. At last an answering click responded to his call.

Laboriously, Shaley tapped out his message of disaster. "Number 17 terribly wrecked forty miles west of Cheyenne. Send hospital train."

A rescue party was organized at Cheyenne and sent speeding to the scene of the wreck, first a work train and then a hospital train with doctors and nurses. Their arrival saved dozens of lives,

Early day transcontinental train enters the loneliness of the Great American desert. Track and telegraph line are the only signs of civilization.

but they were too late to help Frank Shaley. After being assisted down from the pole, he was made as comfortable as possible with his head pillowed on his tool satchel and a pile of cloth. He was dead when the doctors arrived.

Along the railroad, telegraph wires sped messages that kept the remote stations of the lines in touch with the world, as well as the routine orders that kept the freight and passenger trains moving in an orderly, safe manner. When train robbers began to attack the trains on the Western railroads, they knew that cutting off the telegraph service was essential to their crimes, so that they would not be interrupted by sheriffs.

Out on the plains of western Nebraska on the night of September 19, 1877, the Union Pacific depot in the little community of Big Springs was a quiet place. The rest of the town was asleep. The station agent had little to do except listen idly to the occasional bursts of dot-and-dash conversation on the telegraph line. He was reading a book by lamplight, waiting for the night express train to pass by, when the station door burst open with a clatter. Two masked men stood in the entrance to the small wooden depot. They pointed guns at the operator and stepped so close to him that he looked into the barrels of the weapons.

"Put that telegraph out of business!" one of the invaders ordered.

Thinking that he might fool the men, the agent fumbled with the instrument, pretending to disconnect it. He wasn't certain yet what the men wanted. He knew there wasn't enough money in the station safe to make a robbery attempt worth the effort. Perhaps he could tap a warning message on the wire if the men's attention became diverted. The second man, who seemed to know about telegraphy, caught him at this deception and told him to unfasten the wire leading to the sending key. He waved a pistol suggestively to reinforce the order.

"The express is due in five minutes. Hang out your red light and stop it," came the next command.

Soon the train's headlight appeared as a faint glimmer in the darkness and grew larger. Seeing the red lantern hanging at the station, the engineer slowed the train and pulled it to a grinding stop. With a gun at his back, the station agent knocked on the door of the express car. Four more masked men appeared out of the dark around him. When the baggage clerk opened the door, the men jumped aboard with sacks. Ordering the baggage clerk to open the safe, they dumped its contents into the sacks—$60,000 in currency and gold. Either the robbers were extremely lucky to find such a big haul or they had advance knowledge that a large money shipment was being made that night. Some of the robbers held their guns on the agent and the train crew, while two others made their way through the coaches, seizing money and jewelry from the frightened passengers. Then the gang disappeared in the darkness. Moments later the hoofbeats of their galloping horses were heard as they fled.

The station agent ran back into the depot, reconnected the wires, and tapped out word of the holdup to the next station down the railroad line. Brief and daring as it was, the robbery was a reminder of how isolated a corner of the world could be when the telegraph was cut off.

When a gang of train robbers struck the Santa Fe in Oklahoma Territory a few years later under similar circumstances, the telegraph operator tried to send a warning message, was caught at it by one of the gunmen, and shot dead.

In the first years of the twentieth century, the telegraph had a major part in another train robbery, this time because a warning message did get through.

On the night of April 10, 1910, the operator in the station at Tolinas on the Southern Pacific line northeast of San Francisco could hardly believe his own ears as the wire delivered this message: "Wild engine off No. 10 loose on the main line going east. Derail at first opportunity."

Number 10, he knew, was the China-Japan Mail, one of the railroad's fastest passenger trains from San Francisco to Chicago, so named because it carried mail from the Orient that had reached San Francisco by ship. One of his nightly duties was reporting when it passed his station. But if the locomotive was loose and running wild, where was the rest of the train?

Across the track from the station, a freight train stood on a siding, waiting clearance to enter the main line after the express train had passed. The agent ran across to the freight train crew and exclaimed his news. No time could be wasted, because fifteen miles to the east of Tolinas a westbound train was moving along the stretch of single track. If they could not stop the wild locomotive at Tolinas, a head-on crash seemed inevitable.

The runaway engine must be turned off the main track. So, acting on the agent's warning, the freight conductor threw open the switch into the siding where his train stood. Better to have some freight cars smashed than to wreck the oncoming passenger train.

Standing back at a safe distance, the telegraph operator and freight crew stared intently up the track until a headlight appeared. It grew larger still, as the locomotive roared down upon them. The men could see that it was indeed running loose and alone. As it crossed the switch, it swerved onto the siding. With a roar it smashed at high speed into the rear of the freight train, shattering two cars into a pile of wooden debris and scattered freight.

Back at his instrument, the operator reported what had taken

Masked men rob the express car of a Union Pacific train in western Nebraska. Train robbers put the telegraph out of action to prevent a warning being sent.

place. "What happened to the train?" he asked.

Back to him came the story. As the China-Japan Mail rolled eastward from Oakland, its starting point on the eastern side of San Francisco Bay, it came to a strait. No railroad bridge existed there at the time, so the train had to be ferried in sections across the water and reassembled at Benecia. While the sections were being coupled together, two men had sneaked aboard the engine tender in the dark.

Soon after the train had gathered speed east of Benecia, the bandits jumped down from the coal tender into the swaying cab of the locomotive. They were masked. Pointing guns at the engineer and fireman, they took command. "Get out of the way!" one ordered the engineer. The bandit took over the engineer's seat and with obvious knowledge began to run the train himself.

For several minutes the captured train ran through the night. A mile short of Goodyear Station, on a signal from his companion, the bandit engineer brought the train smoothly to a halt. Waving their guns, they ordered the engine crew back to the mail car. The mail clerk realized what was happening and refused the robbers' order to open the door.

"Open up or we'll blast you with dynamite!"

That did it. The clerk changed his mind in a hurry. The door might withstand bullets but not dynamite, and these men clearly were desperate. The door was pulled open. At gunpoint, the engineer and the mail clerk carried the mail sacks from the car up forward to the locomotive. At the same time, the fireman was ordered to disconnect the locomotive from the train.

Once the mail sacks were aboard, the pair of robbers climbed onto the engine and steamed away, leaving the trainful of passengers sitting in the powerless coaches, wondering what had happened.

At Goodyear Station the track crossed a bridge over a slough. An elderly man served as bridge tender. He had been wondering why Number 10 was late. As he watched the locomotive speed across the bridge, he was amazed to see that it hauled no coaches behind it. His surprise turned to fright as the engine stopped a short distance beyond him and the two masked men tossed their loot of mail sacks to the ground. Then, pulling the throttle wide open, the bandits jumped from the locomotive. As the engine gathered speed for its wild unmanned run, the bandits carried the mail sacks down the embankment to the edge of the slough.

By the early 1900's, some points on main line railroads had telephone lines to supplement the telegraph. Goodyear was one. The bridge tender telephoned his news to the train dispatcher at Oakland, who immediately put his warning onto the railroad telegraph line to all points. Tolinas was fourteen miles east of Goodyear, not far for a fast locomotive to travel with its throttle open. But the telephone and telegraph were faster. The speed of their messages, plus the quick reaction of the telegraph operator at Tolinas, prevented a train wreck.

Three months later, the robbers were caught after railroad detectives followed a trail of clues based on the deduction that the men had left Goodyear Station in a small boat. Both were sentenced to forty-five years in prison.

Samuel F. B. Morse could not have conceived when he sent his first message along the wire from Washington to Baltimore that some day his telegraph would play such an essential role in the moving of men and messages in the great spaces of the West. But it was so. Without his telegraph, the opening of the West would have been far slower and less certain.

CHAPTER 12

Men in Green Eyeshades

As telegraph wires spread across the West and the rest of the country in a metallic spider web, the man in the green eyeshade became a familiar and romantic figure. He talked with faraway unseen men in a strange clicking language that few outsiders understood. He belonged to a special fraternity whose members conversed in a private tongue.

The traveler entering a small Western one-story wooden railroad depot opened the door into a drab waiting room whose focal point was a ticket window in one wall. In the center a black potbellied coal stove stood on a sheet of metal to protect the wooden plank floor from live coals, with a stovepipe running up into a hole in the ceiling. Straight-backed wooden benches with iron arm rests lined the walls. A bulletin board was festooned with crudely printed sheets of paper announcing weekend and holiday excursions to nearby resorts, or to exciting distant places like Chicago and Niagara Falls. Chalked on the blackboard were notices of train arrivals—Number 6 eastbound, on time; Number 17, westbound, thirty minutes late.

Presiding behind the brass grill of the window was a man in a green eyeshade. Elastic armbands like garters held his sleeves up from his wrists. He was the ticket agent who doubled as telegrapher. A small black-and-yellow metal sign hanging outside bearing the words WESTERN UNION announced that he served also as the town's sender of birth announcements and death messages to relatives. The railroad station was his kingdom. When a passenger train stopped in obedience to the upraised red semaphore arm outside his bay-windowed throne room, he walked across the platform to the locomotive cab and handed the engineer his telegraphed orders. Less lordly but an essential part of his duties was the business of pulling the baggage cart on its high iron wheels to the open door of the baggage car, there to receive an arriving shipment that might include a bundle of big-city newspapers, a cage of squawking hens, cartons of merchandise for a local store, and the sample trunk of a traveling salesman.

Stepping to the brass grill to purchase a ticket, the traveler did his business with the man in the eyeshade. Although the telegraph instrument in his workroom was clicking while he sold the ticket, the agent ignored it, leaving the traveler to wonder what mysterious conversation was going on over the wire, unattended. Then a certain combination of sounds struck the agent-telegrapher's ear. He strode to the table beneath the wall clock whose swinging brass pendulum behind a glass case ticked off the muted seconds. The telegrapher tapped an acknowledging "I. I." When a train passed the station, he telegraphed that fact to the dispatcher. To the layman it was

Telegraph instrument used by Samuel F.B. Morse prior to 1870

Morse telegraph sending and receiving instruments of 1870

Morse key and sounder used around 1880

fascinating and mysterious. And it went on twenty-four hours a day.

A telegraph office in a big city was like a room abuzz with angry bees. Twenty or more Morse instruments chattered simultaneously, connected to wires that ran from the room to cities hundreds of miles apart.

At each table an operator in eyeshade and shirtsleeves pounded his sending key or translated the staccato flow of incoming sounds into messages on his typewriter. Pneumatic tubes that curved to the ceiling carried the incoming messages from the operators to a central control desk. In an earlier day, all messages were transcribed in longhand. When the typewriter was developed, life became easier for the telegraphers. No longer need they suffer quite so much from "operator's fist," the cramped hand that came from writing down messages steadily and holding the hand in a tight position while tapping the key.

The noise seemed like a jungle of sound. It did not to the telegrapher, however. His ears were trained so delicately that he could pick out the tone of his particular instrument. Some operators placed an empty tobacco tin behind the sounder on the table, to give the incoming dots and dashes a distinctive pitch. For many years the telegraphed messages that kept the business of America running bounced off the stout frock-coated figure of Prince Albert on the front of a thousand bright red tobacco cans.

Although operators at opposite ends of a telegraph circuit could converse only through dots and dashes, they became well acquainted in this manner. Each operator had his own sign, usually one or two initials, with which he ended his messages to show who transmitted them. Q might never meet CG, with whom he worked a wire, or even know his full name, but they felt that they knew each other. A skilled operator had a personal touch on the sending key with delicate variations in his dot and dash strokes. His work could be recognized by other veterans along the wire even before he sent his signature sign. The flow of the telegrapher's code had nuances similar to those of the human voice. During gaps between commercial messages the operators chatted—told jokes in Morse code, exchanged personal opinions and gossip. Despite company rules requiring use of gentlemanly language at all times, operators sometimes dropped in a few swear words.

Not all operators were men. In the days when few women held jobs anywhere else, feminine telegraphers were seen frequently. Having proved their skill at sending and receiving, women often took their places at the telegraphers' tables, even on the railroads. Usually an experienced operator could tell when a woman was

Women telegraph operators sometimes developed romances with male operators at the other end of the circuits, as this drawing of the 1870's, with its cupids and cooing doves, suggests.

sending, by her light touch. Occasionally he was fooled, causing him to change the nature of the jokes he was telling, in embarrassment.

When a man worked one end of the wire and a woman the other, their chatter at times became a flirtation, at a safe distance of two or three hundred miles between the two parties. Occasionally one of these flirtations "took." The operators became so curious about each other that they arranged to meet. Romances developed and then marriages. For many a couple in the late 1800's, the language of love was dots and dashes.

Learning to be a telegrapher often was a trial-and-error business. Training was casual and offhand. A young fellow hanging around a railroad office was allowed to try his ear and hand at copying messages that passed along the wire, until he could distinguish dots and dashes in the flow with some confidence. After practicing on a key, he was permitted to send a few simple messages with a hesitating hand. The receiving operator knew immediately that he was handling a beginner and usually was sympathetic. When the apprentice was judged to be able to fend for himself, barely, he might be sent to "solo" as a night operator in a small station that had little traffic. Much as he tried to hide his inexperience, often he had to signal "bk . . . bk"—break, break—and ask the sender to repeat words he had missed. Embarrassing as this was, it was better than trying to guess at the missing words and filling in wrong ones.

In the early days of the telegraph, before commercial distribution of electricity was common, tending the batteries that generated power for the line was one of the telegrapher's duties. The Western Union handbook of rules for operators issued in 1860 gave instructions on how to take care of the battery:

> Clean jar, set copper in jar, spreading it as wide as jar will admit. Fill the pocket with pulverized vitriol and hang on the edge of the jar where the copper is open. Having filled porous cup with soft or rain water, sufficiently full as not to overflow when the zincs are inserted, put the zinc into the porous cup, having first placed the cup inside the copper. Pour soft water slowly through the vitriol in the pocket, filling up with the pulverized vitriol as it is reduced by the flow of water. No vitriol should be allowed to be dropped into either the jar or the porous cup.

Once he mastered the mysteries of the battery, the operator would start work on the wire. If he didn't, no messages could move.

During heavy lightning storms operators were instructed to cut off the main telegraph wires from their instruments as a safety precaution. Many placed insulators under the legs of their tables for the same reason.

Reminding us of how many Americans could not read or write

Replenishing the batteries that produced power to make the circuits operate was a duty of pioneer telegraph operators.

in those days, the Western Union handbook also contained this instruction for the handling of money: "Vouchers receipted by 'his mark' must be witnessed by some person other than the one disbursing the money."

These Morse telegraphers liked to show off their skill and speed in sending contests. Hundreds of telegraphers gathered in a hall to watch the stars perform. On stage were several tables with telegraph keys. Each contestant in turn was handed an excerpt from a speech. He tapped this out on his key while the audience listened in silence, as if to the performance of a concert violinist. They could "read" the flow of dots and dashes like a book, their ears breaking down the sounds into letters and their minds translating them into words. Speed and accuracy—those were the goals. When an especially swift operator sent at a pace of fifty words a minute, his wrist made 750 separate motions in that sixty seconds. If a contestant did an unusually brilliant job of transmitting, his performance was greeted with a standing ovation from the audience. This was personal craftsmanship developed to the utmost, when automation was still far in the future.

"Talking" on a Morse wire was done in a kind of shorthand. The sender used abbreviations to save time. Instead of tapping, "Where have you been?" the operator merely sent, "Wr u ben?" If he called the operator at the other end of the circuit and received no answer, he messaged, "Wy dut u ans?"—"Why don't you answer?" Hundreds of these shortcuts were included in the "Phillips Code," which operators memorized. They used this code only in wire conversation; all words in formal messages were spelled out in full.

Old-time newspapermen who learned their trade when Morse wires still were in existence found many of these telegraphers' shortcuts so handy that they incorporated them into their note-taking on assignments and continue to use them. The writer of this book, for example, always writes HX for Chicago, NX for New York, Scotus for Supreme Court, t for the, n for in, tt for that, and trou for trousers. For some reason, the abbreviation for Honolulu is 99. When an old-time telegrapher tapped, "99warding," he meant that he was going to Honolulu, though not many of them ever had the opportunity to do so. When the time came for a ten-minute rest period on the wire, one operator would tell the other, "Take ten." This term came into popular usage, especially around the motion picture studios.

Morse operators came to think in sounds. Usually they were excellent spellers. If one of them had trouble remembering how a word was spelled, he would tap it out with a fingertip on the arm of a chair, or perhaps on his teeth, until he recognized the sounds as being right. There are tales, too, about telegraph operators in card games out West who tipped off their partners about the contents of their hands by tapping a finger on the card table like a nervous habit while thinking—a dishonest trick that could lead to more than angry words if the perpetrator was caught.

News dispatches were sent across the country by Morse code over the leased wires of the press associations, The Associated Press and United Press. An operator sat in the office of each newspaper on the circuit, copying the dispatches sent from the central press association office. The words flowed at about thirty a minute. The receiving operator handed the copied messages to the editor, to be put into type. This operator was always the first man in his city to learn of big news breaking around the world. In those days before television and even before radio, the World Series was by far the greatest American sporting event. When the games started, the press associations quit sending all other news and telegraphed each game, pitch by pitch, to their client newspapers. Many of these papers posted scoreboards and charts in the streets outside their offices, where hundreds of fans gathered to watch the action.

Let us say that the game was being played in St. Louis. The operator in the press box would send "strike" or "ball" or "out fly left." The operators in newspaper offices in Los Angeles, Seattle, and New Orleans copied this word simultaneously. If the game became tense, tied in the bottom of the ninth inning with bases full, each operator became the hero of his office. Only he could interpret the dots and dashes. Others in the office crowded around his table waiting for the telltale clicks; then his pronouncement of "Strike three!" or "Home run!"

During the pioneer days of the West when justice was swift and often harsh, the telegraph pole occasionally was a handy gallows for quick execution of a murderer or horse thief in a remote region. A judge passed sentence of death and the criminal was taken unceremoniously by the sheriff to a pole near the shack that served as a courtroom, a rope was swung over the crossbar, and soon the offender's body dangling from the pole was a grim warning to others against such conduct. At times, angry vigilantes who caught an offender didn't bother about the formality of a trial but headed direct for the nearest pole with their captive. Usually the right man was hanged, but if the victim wasn't really the villain, he was in no position to argue. Lawyers with their courtroom maneu-

A vigilante court delivers quick justice. While the judge listens to the offender's plea for mercy, a man strings a hangman's noose from a nearby telegraph pole.

vers, delaying actions, and technical challenges of the law had no place in most frontier trials.

If the telegraph was the means for some Western executions, it also was the method by which other men were saved from unjust hanging.

Early in the 1850's, a young Indian confessed to the murder of a mail carrier on the Santa Fe trail. He was being held in the prison at Jefferson City, Missouri, awaiting execution, when it was discovered that he had confessed to protect his father, the real murderer. He was to die on March 14, 1851. Too little time remained for a written appeal on his behalf to be sent to President Millard Fillmore in Washington in the normal way, so the appeal was sent by telegraph on March 13, apparently the first use of the wire for this purpose.

The appeal reached President Fillmore in the White House that night. He could not reply until the next morning, however, because the telegraph office was closed! Early on the fourteenth, the President sent a message: "The Marshal of the State of Missouri is hereby directed to postpone the execution of the Indian See-see-sah-ma until Friday, the 18th of April."

Time was growing short. No direct telegraph line yet existed from Washington to St. Louis. So the President dispatched his life-saving message over three routes; in each case the message was relayed along the wires of four different companies. One copy was ferried across the Ohio River at Cairo, Illinois, in a rowboat.

The first of his telegrams reached the marshal's office in St. Louis just before 10:00 A.M. The execution was scheduled for noon. The marshal forwarded the reprieve by wire to his deputy at the Jefferson City prison, where a crowd of hundreds had assembled to watch. Holding the telegram in his hand, the deputy marshal informed See-see-sah-ma that his life was saved. He later was set free after the appeal papers were signed.

An even stranger tale comes down to us from the early days in Arizona, when the few inhabitants lived in scattered settlements

Western outlaw gets swift, fatal justice at the end of a rope suspended from a telegraph pole.

with skimpy communications. In a remote county a man condemned for murder was on the eve of his execution when another man confessed that he had committed the crime. However, the sheriff had to obtain permission from the governor to call off the execution of the condemned man so that the matter could be thrashed out legally. The sheriff rode his horse to the nearest telegraph station and sent a message to the governor for a reprieve.

It was late at night when the sheriff's telegram reached the state capital. The governor had been at a party and had celebrated more than he should have, to the point that efforts to rouse him were in vain. The other official with power to act was out of town. So the telegraph operator took it upon himself to save the man's life. He sent this message to the sheriff:

> Reprieve is granted to (prisoner's name) for ten days. Regular papers go forward in the morning.
>
> ——————————, Governor, per ———————— (operator's name), Acting Governor, pro tem.

The sheriff's deputies actually had the hangman's rope ready when the sheriff galloped up to the gallows with the telegram. About the same hour, the governor saw the sheriff's telegraphed request, realized what time it was, and moaned, "I have let an innocent man die."

"No, you didn't, sir," the telegrapher told him. "I took your place for fifteen minutes and saved him."

As the West grew more heavily populated, telegraph lines were strung to many out-of-the-way towns. Postal Telegraph came into business as a rival system against Western Union, although its service was concentrated mostly in larger cities. On a dreadful morning when the twentieth century was new, one of the country's most spectacular disasters put telegraph lines and operators to an historic test.

At 5:15 A.M. on April 18, 1906, two men were on duty in the Associated Press office in San Francisco, a Morse telegrapher and an editor. CRASH! The building rocked, plaster fell from the ceiling, a clock tumbled to the floor. Window glass tinkled on the desks.

"An earthquake!" exclaimed the editor. "Here, send a bulletin." The operator tried to pound out the words. No sounds came. The circuit had been destroyed.

He tried another circuit. Still, only silence. San Francisco had been hit by a massive disaster and no way existed to tell the news.

"Let's carry the story over to Oakland and send it from there," the operator suggested. He and another operator made their way through the debris in the streets to the waterfront and rode the ferry across the bay to Oakland. Their attempt was useless; every telegraph line in Oakland also was smashed.

Back to San Francisco they came. Hours passed. Still the world did not know about the tragedy that had struck San Francisco. The AP men went to the commercial office of Postal Telegraph, where they found the staff tinkering frantically with their wires and keys, trying to get a circuit up.

After some time a Postal operator heard a muted clicking in response to his calls on the wire and announced, "I've got a circuit going to Chicago." He transmitted the bulletin that the AP men had been trying so long to send. Before many more words could be sent, however, fire swept close to the building and the fragile circuit, San Francisco's one lifeline to the world, broke down. The AP men made their way along the crumbled street to Western Union, hoping for better luck there. Western Union got a circuit working, but hardly had the operator begun sending than police arrived with word that the Western Union building was to be blown up in an attempt to stop the fire from spreading.

Thus it went for hours, one patchwork circuit after another. The AP men poured out the words describing the tragic results of the earthquake and the fire that followed, and the Morse wires sent them out. The world was shocked by the stories it read. In twenty-four hours, the Associated Press men sent 21,000 words of description by Morse wire, depicting the massive destruction that shattered the city on the Golden Gate. Soon help and emergency supplies were arriving from many directions. The telegraph had proven what a vital tool of civilization it was.

CHAPTER 13

"Mr. Watson, Come Here"

Fifteen years after the transcontinental telegraph line was completed, wires that had "talked" with dots and dashes began to talk, literally. Instead of transmitting coded messages, they carried the human voice. Alexander Graham Bell invented the telephone, a creation even more dramatic than the telegraph in the impact on persons' lives. Very few teen-age girls, for example, ever find need for the telegraph, but many consider the telephone indispensable for their survival.

Bell was a young voice teacher from Scotland. He had moved to the United States and made an uncertain living teaching the deaf, as his father did in England. Along with knowledge of the human ear, he had a flair for electricity. In Boston during the early 1870's he carried out experiments searching for a way to transmit the human voice by wire. He had been working on a plan to improve the telegraph by sending several messages on a single wire simultaneously, when his experiments were diverted toward voice transmission, instead.

A week after his twenty-ninth birthday, he and his assistant, Thomas Watson, strung a wire from Bell's room on the third floor of their Boston rooming house to Watson's on the floor below. They wanted to test a new device Bell had developed. A few months earlier the two had succeeded in transmitting voice tones along two hundred feet of wire, but the results were unintelligible—just sounds whose meaning could not be deciphered by the listener.

On Bell's table was a curious-looking instrument, crudely made,

Artist shows Alexander Graham Bell with his first telephone in attic workshop at 109 Court Street, Boston. It was this instrument that first sent human voice sounds over a wire.

When Alexander Graham Bell spilled some acid near the telephone transmitter he was about to test, he called to his assistant, "Mr. Watson, come here; I want you." This was the first telephone message.

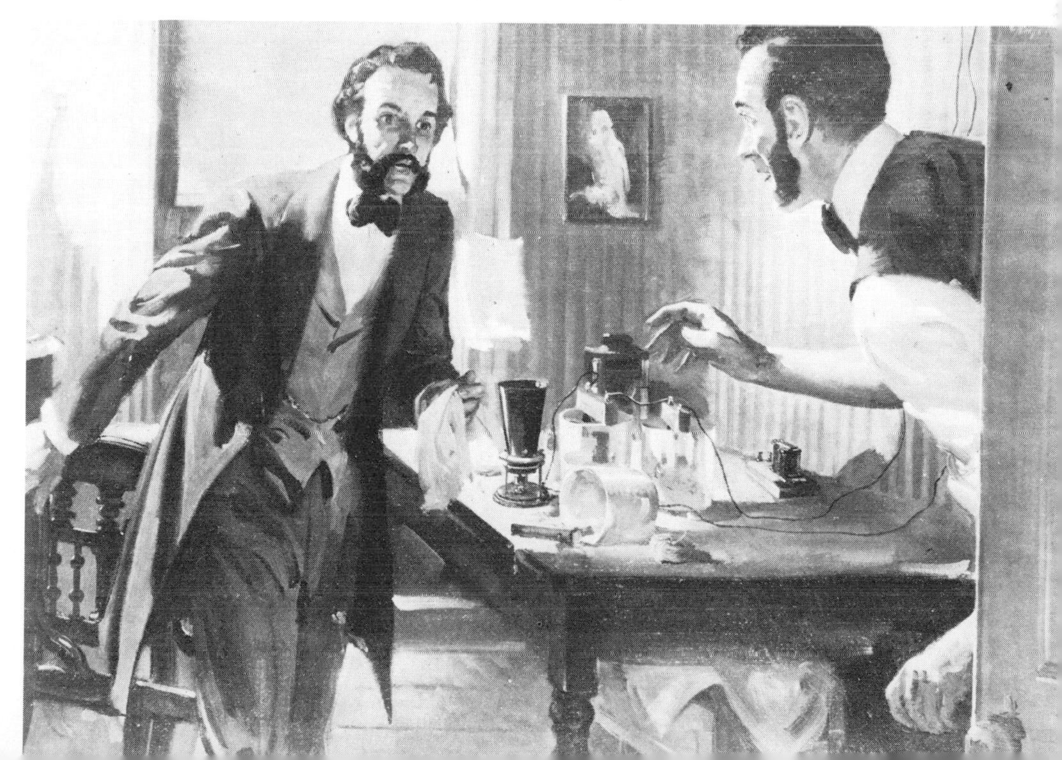

that somewhat resembled a narrow iron bowl with an open top, resting in a little wooden stand.

Bell spoke into the instrument, "Mr. Watson, come here; I want you."

In his room downstairs, Watson's face lit up with excitement. He ran up the stairs and told Bell, "I heard you. I could hear what you said!"

The moment was historic. This was the first complete sentence ever transmitted by telephone. The date was March 10, 1876. The year is easy to remember, because it was the one hundredth anniversary of American independence.

Although Bell is officially credited with inventing the telephone, others were working on the same project unknown to him. A short time before his famous telephone message to Watson, Bell's attorneys filed an application at the Patent Office in Washington for a patent on his telephone invention. A few hours later in the same day, Elisha Gray filed in the same office a document called a caveat, which in effect was a notice that he was developing a telephone. The coincidence of two inventors making known officially their development of a telephone on the same day was remarkable, to say the least—one of history's flukes.

This led to one of the great patent fights in the history of American invention. Hundreds of millions of dollars in potential earnings from the patent were at stake. The courts decided finally that Bell was truly the inventor of the telephone.

In the centennial summer of 1876, however, Alexander Graham Bell was more concerned with convincing the world of the value of his invention than he was with a court battle. He demonstrated the telephone at the Centennial Exposition at Philadelphia that summer to the applause of the country's leading scientists. Making the public understand the value of his discovery was not so easy.

Like Samuel F. B. Morse with his telegraph thirty years earlier, Bell found many who scoffed at his invention. Some newspaper editors were impressed while others belittled it. Because part of Bell's work was done at Salem, Massachusetts, site of the colonial witchcraft trials, one editor called the telephone "Salem witchcraft." To the New York *Herald*, the effect was "weird and almost supernatural." The Boston *Times* in a joking tone said, "A fellow can now court his girl in China as well as in East Boston; but the most serious aspect of this invention is the awful and irresponsible power it will give to the average mother-in-law, who will be able to send her voice around the habitable globe."

Perhaps this doubt shouldn't seem surprising. Bell's demonstration at the Centennial Exposition was not always reliable, producing a garbled sound at times. Users of the first telephones had to shout into them, and their static-ridden voices came faintly through the instrument to the listener. Many who were offered the opportunity to talk felt foolish, yelling into a strange instrument and asking, "Can you hear me?" The world had trouble in 1844 believing that human thought could be carried from one place to another by an

Artist depicts the moment at the Philadelphia Centennial Exposition of 1876 when youthful Alexander Graham Bell demonstrated his invention to Dom Pedro, Emperor of Brazil.

Central telephone office in 1879, staffed entirely by boy operators. Male operators were common on night shifts until after 1900, because it was considered unwise for girls to hold jobs that took them away from home after dark.

electrically charged telegraph wire. Now many persons had difficulty in recognizing the potential of Bell's invention.

Although both used wires strung from poles, the telephone and telegraph had different principles. In the telegraph, surges of current of varying duration sent through the wire from one point to another, with moments of silence between, formed the message. The basic theory of Bell's telephone was almost as simple. The sound waves of the human voice strike a thin electrically charged diaphragm in the telephone transmitter, making it vibrate. This creates fluctuations in the current flowing through the wire to which the telephone is connected, making an electrical "pattern" of the speaker's voice. At the other end of the line, these fluctuations in the current strike a diaphragm in the receiver, causing it to vibrate. This diaphragm transfers the fluctuating electrical current variations back into sound waves that are received in the listener's ear as human speech.

At first, Bell used existing telegraph wires to transmit his telephone calls experimentally between cities. The first telephone calls between Boston and New York were made in 1877 over railroad telegraph wires. That same spring, the first local telephone line was opened in Boston between a businessman's home and his office. Telephone lines at that time ran directly from one point to another, so that only the parties at either end could use them. Soon it was apparent that a central place must be created to which all the local wires ran, so that a call from one point could be directed to any other telephone in the company's system. Thus the telephone exchange was born.

The appearance of the telephone instrument changed as Bell's system developed. Early commercial phones were boxes on a table that served both as transmitters and receivers. Next the phone box was hung on the wall with two instruments dangling from it. The user held one to his mouth for talking, the other to his ear for listening. From this evolved the wooden-box wall telephone that was a familiar sight in thousands of American homes in the early 1900's. It had a receiver on a hook, a mouthpiece projecting toward the user, and a small crank handle at the side. To place a call, the user lifted the receiver from the hook, turned the crank to summon the central operator, and when she answered, "Number, please?" told her whom he wanted to reach.

Women telephone operators on duty at early day telephone central exchange office

Telephone operators at central switchboard in Milwaukee, Wisconsin, 1883

Long distance switchboard in New York, about 1892. Notice the mixture of men and women operators.

Central telephone office operator with equipment, 1880

Telephone service of this primitive nature developed in cities on and near the Eastern seaboard in the 1880's. To the people in the sparsely populated West, however, the telephone for some time was something heard about but never seen. If a person wanted to send a message by wire, the telegraph still was the only way.

Gradually telephone lines were extended from city to city in the East. Cities built their own telephone systems for local use but often had no connection with phone companies in other cities. These local exchanges were islands of phone operation waiting to be hooked together by long distance wires. The same kind of local service developed in San Francisco, Denver, Los Angeles, and a few other Western cities.

Line of telephone poles on West Street which brought the Chicago-New York circuits into New York City in the early 1900's

That old bugaboo of great distances in the West hampered development of telephone service there, just as it had done with the telegraph. Since the cities were rather small, and the distances between them large, slight economic incentive existed to build long distance lines that might not carry enough phone calls to pay their cost. Sixteen years passed from the time Bell spoke his magic words into his experimental telephone in 1876 until the first line was opened from New York to Chicago in 1892.

Use of the long distance phone was slow in coming, even then. One night in 1898 during the Spanish-American War, the telephone rang in the White House. When the night watchman answered, a newspaperman calling from Chicago asked to speak to President McKinley. The idea of disturbing the President at night for a telephone call startled the watchman, but he was so impressed by the long distance call that he agreed. The sleepy President came to the phone in his night clothes and was delighted to hear that Admiral George Dewey had defeated the Spanish fleet at Manila. Compare that call with the massive twenty-four hour communications, including a hot line to Moscow, that link the White House with all corners of the world today!

The huge land gap between the Midwest and the Pacific Coast, across plains and mountains, challenged the telephone builders just as it had the telegraph men. This time the problem was financial as well as physical. The cross-country telegraph builders had a subsidy from the federal government to support them, but the telephone builders did not. The shrewd businessmen running the telephone industry weren't going to spend hundreds of thousands of dollars for telephone lines across the West until they were sure they could earn sufficient money from their venture. There were severe technical problems, too. Until equipment became better developed, the possibility of carrying a voice more than three thousand miles from New York to San Francisco was uncertain, at best. Only after the Western cities had grown substantially in size did the telephone company undertake the task.

Building the lines was only part of the problem. Keeping them operating was almost as difficult. The poles carrying the wire stretched mile after mile through unpopulated western mountain regions, far from any community and remote from roads. They climbed through dense forests and up mountain gorges above the timber line where no trees stood and storms swept unhampered at elevations of twelve thousand feet and higher, snapping poles

Often working in foul weather, the lineman was an essential figure in maintaining service on telegraph and telephone circuits, before they were put underground.

Below: *An early service car operated by the Pacific Telephone and Telegraph Co., in San Jose, California, in 1912*

and blowing down the wire. Telephone linemen had to hike, or when possible ride horseback, along the lines day and night, in sub-zero cold and hot summer, hunting the breaks in the circuit and climbing the poles to repair them. Sleet storms were among the worst hazards. Ice gathered on the wires to a thickness of several inches, bending them to the ground. A repairman carried a load of about fifty pounds—wire, test set and spiked climbing irons that he attached to his boots to help him up the poles. In winter he used skis to reach the trouble spots. One break in the wire could halt service for hundreds of miles, so the lineman had to keep moving no matter what the obstacles.

And there were unexpected obstacles, indeed. The buffalo that knocked down poles while using them to scratch their backs were a menace to the telegraph builders; for the telephone men it was the bears. In the mountains these powerful furry creatures gnawed the poles to the ground, apparently thinking that the hum of the wires was the buzz of bees. A lineman in the mountains located the source of one break in the service—a golden eagle had become caught in the wires. The lineman climbed a pole to remove the bird's body and was attacked by the bird's mate, which clawed and bit him furiously.

Early in the twentieth century, the practice of burying well-insulated telephone wires underground to replace the festoons of wires hanging on high poles was begun. This not only looked better but reduced the danger of broken wires. Many years passed, however, before underground wires became commonplace in thinly populated areas. Visitors to ranch regions in the West still see lines of poles carrying a single telephone wire along a dirt road to a ranch house set in a clump of trees far from the highway. Even when the wires are buried, they are not entirely free from interruption by curious animals. Telephone men in a New Mexican town reported trouble from a gopher with an odd sense of color. The animal sank its sharp teeth into an underground phone cable containing wires of eleven colors. For some reason, it nibbled only on the yellow wires. But what a feast it had!

As telephone lines reached more places in the West, the blessings they could bring to isolated farms became apparent. Among those

New York street scene in the 1880's, Broadway from Maiden Lane, in the days before telephone wires were placed underground. Some poles had fifty crossarms. It is interesting to compare the photograph and the drawing.

who discovered this were the fruit farmers in Colorado. The spring frosts that frequently laid a killing white hand on the budding fruit trees were a menace, often causing the loss of many thousands of dollars to the farmers. In 1909 these growers tried something new. They bought hundreds of smudge pots, small stoves with long chimneys that poured out clouds of black smoke from the oil burned in them. When placed in the orchards, the pots produced a sooty, heavy pall, preventing frost from forming on the trees.

That spring when the smudge pots had been placed in position, the United States Weather Bureau reported that a frost was probable. Great peril threatened the apple trees, then in pink blossom. The message went out by telephone to all the farmers who had phone service, "Be ready to light your smudge pots in half an hour."

The farmers in turn called friends in nearby towns for aid. By

Maze of wires in Pratt, Kansas, in 1909, was a hint of the impact of the telephone on the Southwest.

wagon and by horseback help came—hundreds of men to assist the fruit growers in firing up the smudge pots.

Soon the Weather Bureau telephoned again, "The thermometer registers twenty-nine. Light the fires!"

The emergency crews ran up and down the rows of fruit trees, touching fire to the smudge pots, until a filthy but life-saving cloud filled the air from the ground to a level above the treetops. This kept the temperature above the freezing mark. A crop worth three million dollars was saved, a rescue that would have been impossible without the telephone.

The nineteenth century turned into the twentieth century, and still the idea of a transcontinental phone call from New York to San Francisco remained a dream.

When the cross-country telephone line finally was built, it followed much the same route Creighton and the other telegraph builders had used more than fifty years earlier. Crews installed a line of poles, stronger and larger than the original telegraph standards, from Omaha to Denver in 1911. Two years later in 1913 the line reached Salt Lake City and hooked into the local phone system of the Mormon city.

At last on January 25, 1915, the long-anticipated day came. A line of 130,000 poles extended across the United States from New York to San Francisco, carrying the telephone lines. In a room on the fifteenth floor of the American Telephone and Telegraph Company building in New York, a crowd gathered around Alexander Graham Bell. The man who had invented the telephone was sixty-seven years old, a distinguished and honored figure with a full white beard. A similar crowd stood around Watson in San Francisco. For the occasion the telephone circuit had been extended south from New York to Washington, D.C., and to Jeykl Island, Georgia, where the president of the telephone company, Theodore N. Vail, listened. In Washington, the telephone of President Woodrow Wilson was connected to the circuit. Altogether, the circuit was 4,300 miles long.

Precisely at 4:30 P.M. on a signal from the technicians Bell picked up the phone and said, "Mr. Watson, are you there?"

"Yes," replied Watson. Hundreds of listeners in New York, San

Early telephone pole line construction, about 1890

Erecting a telephone pole, around 1900

It took until January 25, 1915, before the transcontinental telephone line was completed. Thomas A. Watson, who answered Alexander Graham Bell's historic first telephone call, talks from San Francisco to Bell in New York on the great day.

Francisco, and Washington broke into applause. These were the first human voices heard from coast to coast, the same voices that had spoken in the original phone call thirty-nine years earlier. During that time tremendous improvements had been made in telephone techniques, and the instruments Bell and Watson used bore scant resemblance to the first ones.

Then Bell stepped to an exact replica of the original telephone instrument he had used in 1876, which was hooked into the San Francisco circuit. Again he spoke the historic words, "Mr. Watson, come here; I need you."

Watson replied, "It would take a week for me to get to you this time."

President Wilson, who had been listening in Washington, came onto the line, congratulated Bell and Watson, and declared that the moment was a memorable one in American history. Indeed it was. The Atlantic and Pacific coasts were fastened with a new bond. First the telegraph, then the transcontinental railroad and the Panama Canal, and now the telephone; the sinews of communication that bound the United States together into a common whole were knotted so strongly that they never could be broken.

CHAPTER 14

Farewell, Brass Pounders

Today the hand-operated Morse telegraph that brought the West into instant contact with the rest of the country is only a memory. The "brass pounders" who sent the dot-and-dash messages are a vanished breed. Like the crank telephone on the wall, the horse and buggy, the lamplighter, the steam locomotive, and other once familiar sights in the American past, they have been replaced by newer, better methods. Messages from one part of the United States to another are being sent today by electronic systems beyond the most fantastic imaginings of the old-timers. Instead of being transmitted along rickety land lines at thirty words a minute, they flash from one point on earth to another by relay from a space satellite.

During the first years of the twentieth century, inventors developed ways to improve the telegraph system. Methods were found to send several messages over the same wire at the same time, greatly increasing the capacity of each circuit. Instead of tapping out dots and dashes, the operator typed the message on a keyboard resembling a typewriter. When he struck a key, a certain combination of holes was punched in a strip of paper tape running through the transmitter. Pins slipped into the punched holes, creating electric impulses that were transmitted along the wire.

At the other end the electric impulses were translated back into letters and numbers. These appeared on a long thin strip of gummed tape emerging from the receiving machine. The operator pasted the tape on a telegraph blank, which was delivered to the person addressed. This tape could pile up for awhile if the receiving

The quadraplex telegraph system in 1884, allowing transmission of four messages on a single wire. Drawing shows how messages traveled from one office to another.

operator was absent. No longer was the intimate person to-person contact between operators necessary. From the moment that this tape machine went into operation, the days of the old Morse operators were numbered.

Next came the faster, easier teleprinter. The operators called it a "page printer" because the telegraphed words emerged from the machine typed in lines on a continuous broad strip of paper that rolled out from the top of the teleprinter and could be ripped off the machine in page lengths.

About thirty years ago an even more exciting development went into operation. Instead of being sent on wires or through underground cables, telegrams now were sent by a radio beam system, or microwave. Thousands of telegraph poles laden with a mass of wires were replaced by tall, spidery metal towers placed thirty miles apart. In microwave transmission, messages are sent by an antenna in a narrow beam aimed at one of these towers. There

A wireless station, the beginning of the end for the brass pounders

it is relayed automatically to the next tower, and thus across the country. With this system, two thousand telegrams can be sent simultaneously over a single radio beam system. Not only is this much faster but it is not subject to the whims of weather. Those traditional foes of the telegrapher, ice and electrical storms, winds and fallen trees, have no effect on the microwave. This system of transmission is used by telephone companies and private communications companies as well as the company whose name is so traditionally linked to the telegraph in the United States, Western Union. Even more fascinating is the way messages transmitted from earth are delivered to astronauts hundreds of miles aloft by the teleprinters in their space ships.

How far we have come since the pioneer days in the hot desert of the Imperial Valley in California! No wood was available for telegraph poles in that remote area of sand and salt flats, so telegraph messages were sent through the strand of a barbed wire fence that ran along the single railroad track. Frequently when the circuit broke down, investigators found that perspiring crews working on the tracks had hung their shirts on the fence, disrupting the feeble electrical flow.

Western Union's business has changed immeasurably since the era of Morse telegrams. Few family messages are sent by telegram now because making a long distance telephone call by direct dialing is so easy. But Western Union initiated other services, even a singing telegram. If a person wanted to send a birthday telegram to a friend and was willing to pay the extra price, the delivery boy sang the message to the recipient when he delivered it. And it was often difficult to tell who was more embarrassed by such musical efforts, the delivery boy or the person being serenaded.

Candygrams were introduced, allowing boxes of delicious chocolates to be ordered and charged by phone, and delivered promptly in distant places. Instead of candy, small dolls could be ordered by Dolly Grams. Flowers, too, could be sent by wire.

A recent telegram service is called Mailgram. A person in New York wishes to send such a message to a friend in Denver. She calls Western Union and gives her message. This is typed into the Western Union computer and delivered to a receiving machine in a post office near the home of the person addressed. There it is placed in a special Mailgram envelope and delivered by the postman the next day.

A large proportion of the telegrams Western Union transmits today are carried on direct private wires from the office of one commercial firm to another, in contrast to the old days when the sending firm dispatched its message by runner to a local telegraph office. At the other end, a Western Union messenger boy in uniform carried it from the receiving telegraph office to its destination, either on foot or on a bicycle.

A modern Mailgram, a recent telegraphic service

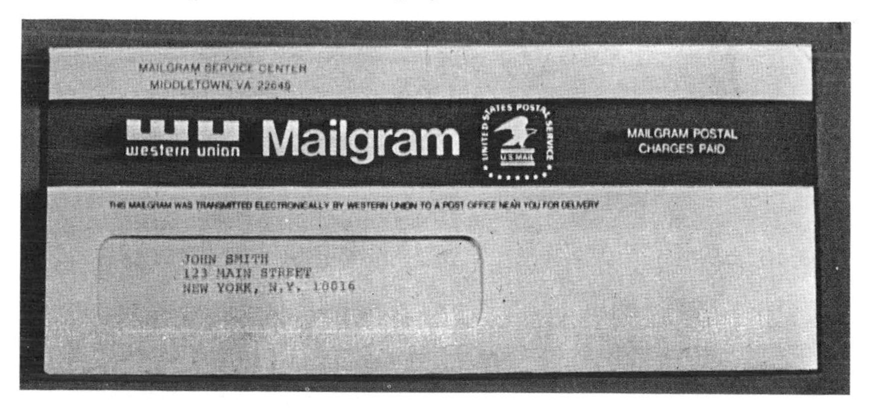

The Morse sending key and receiving instrument in later years, before it was replaced by teletype machines

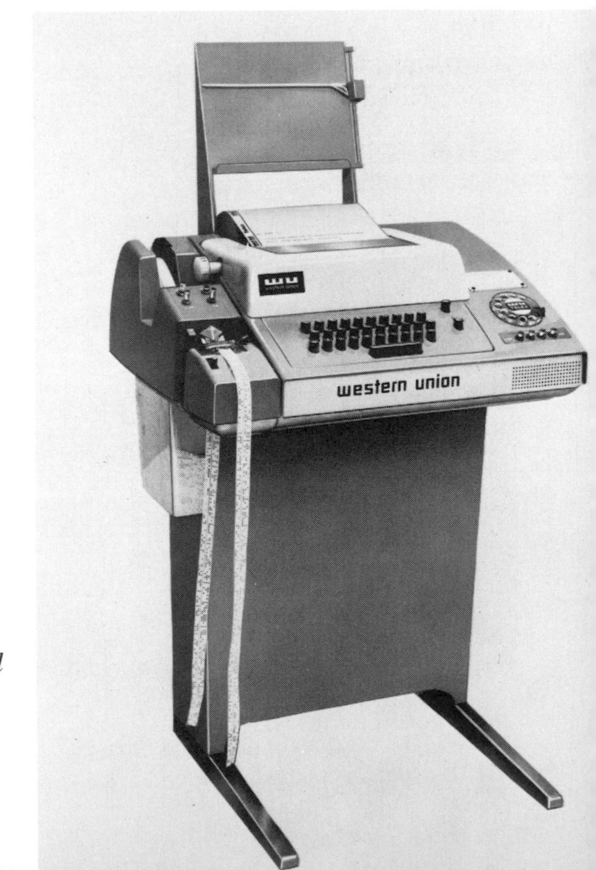

A Telex terminal

Now everything is mechanized. A secretary sitting at a keyboard in the originating company's office punches the messages into perforated tape. She dials the telephone number of the destination in Los Angeles, using a dial on her TWX set. When the connection is made, a signal tells her to go ahead. No one need be present at the receiving machine. She pushes a button and starts the tape running automatically through the sending unit, the various combinations of punched holes creating electrical impulses that represent letters. The words appear on a continuous roll of paper in the Los Angeles firm's office, to be removed at the recipient's convenience.

More marvelous things are happening, so fast that keeping up with them is difficult. Space satellites which catch sound beams aimed at them from earth and bounce the signals back to receiving stations in other cities and countries are supplementing microwave relay systems on earth, and to some degree replacing them. For many years after the old Morse telegraph circuits went out of use, the giant press associations that transmit news stories to newspapers and radio and television stations throughout the country sent and received their stories on teleprinters. For sending, these machines that once seemed so marvelous to the Morse operators they replaced are in turn being eliminated. News stories are prepared on video display terminals by reporters and editors, then stored in a computer. When he is ready to transmit the story, the editor calls it from the computer and it is sent onto the news wire automatically.

Truly, it is a different world of communications today—a world in which messages move from point to point at tremendous speeds, sent and received by computers and other electronic devices. After the experience of sitting in our own living rooms and listening to men talk as they walked on the moon, we have come to take almost any form of communication for granted. Watching these marvels at work, it is easy for us to forget the men who struggled to build the first telegraph across the country and fought the Indians to keep it running. This should not happen. Their story of ingenuity and bravery is such a basic if little-heralded part of our history that it needs to be kept alive. Without the pioneers who ran the wires across the West, today's wonders of instant communication around the world could not exist.

The Plains, heart of the country through which the telegraph line was
Archives of Reverend Eugene Buechel, S.J., St. Francis Mission, St. Francis,

built in 1861. This picture of St. Elizabeth's Church is from the Photo South Dakota (Little Sioux).

Acknowledgments and Bibliography

I wish to express appreciation for help in the preparation of this book especially to Mike Byrne, director of public relations and public information at Creighton University, Omaha, Nebraska, who supplied a copy of the typed manuscript diary of Charles Brown and material about Edward Creighton from the university collection.

My gratitude also goes to the South Bend Public Library for use of its nineteenth-century magazine files, to Julius Ivancsics of the *South Bend Tribune* for photographic assistance, and to the public relations staffs of the Union Pacific Railroad, Western Union, and the American Telephone and Telegraph Company for their contributions of illustrations.

Among the source materials I found most helpful were:

Wiring a Continent, by Robert L. Thompson (Princeton University Press, 1947)

Old Wires and New Waves, by Alvin F. Harlow (Appleton, 1936)

The American Leonardo, A Life of Samuel F. B. Morse, by Carleton Mabee (Alfred A. Knopf, 1943)

The Western Union Telegraph Co. *Rules, Regulations and Instructions* (1866)

The City of the Saints, by Richard F. Burton, edited by Fawn M. Brodie. Originally published 1861 in London. (Alfred A. Knopf, 1963)

Tent Life in Siberia, by George Kennan (G. P. Putnam's Sons. Originally published 1870, expanded edition reissued 1910)

The Indian War of 1864, by Captain Eugene F. Ware, originally published 1911. (St. Martin's Press, 1960)

American Indian Frontier, by William Christie Macleod (Alfred A. Knopf, 1928)

The Fighting Cheyennes, by George Bird Grinnell (University of Oklahoma Press. Original copyright 1915, reprinted 1956)

Moguls and Iron Men, by James McCague (Harper and Row, 1964)

Great Train Robberies of the West, by Eugene B. Block (Coward and McCann, 1959)

Beginnings of Telephony, by Frederick Leland Rhodes (Harper and Bros., 1929)

The New York Times microfilm files, especially during the 1860's

Index

Alaska, 99, 102-103, 105, 112
Associated Press, 141, 144-145
Atlantic cable, 96-97, 109, 117

Baltimore, Maryland, 14 16, 18
Barnum, P. T., 21
Bee, Frederick A., 32
Bell, Alexander Graham, 146-151, 160-163
Black Kettle, 81
Boston, Massachusetts, 150
British Columbia, 104, 111-112
Brown, Charles, 41, 43, 51; diary, 52 63
Brownsville, Nebraska Territory, 23-26
Buffalo, 7-9, 64 65
Bulkley, Colonel Charles S., 100-102, 104
Butterfield Overland Stage, 2, 31-32

California, 2, 3-4, 26, 27-30, 44-45, 81, 166
Carnegie, Andrew, 22
Carson City, Nevada, 32, 34, 36-37, 39, 66
Cheyenne, Wyoming, 124, 126
Chicago, Illinois, 23, 27, 122, 155
Clay, Henry, 16-17
Cody, William "Buffalo Bill," 4
Collins, Perry McD., 98-99
Congress, 14-15, 30, 44
Creighton, Edward, 43, 47-50, 55, 57, 61-63, 64-65, 68-71, 73, 76, 78, 160
Creighton University, 76

Denver, Colorado, 85, 89, 90-93, 153, 160, 167

Edison, Thomas, A., 22-23
Ellsworth, Annie, 17
El Paso, Texas, 31

Field, Cyrus W., 96-97, 99-100, 109, 117
Fillmore, President Millard, 142-143
Fort Churchill, Nevada, 39-40, 45, 50, 65
Fort Kearney, Nebraska Territory, 45, 49, 51, 55, 82-83, 90
Fort Laramie, Wyoming Territory, 49, 60, 64, 90
Frelinghuysen, Theodore, 16

Gamble, James, 71-73, 74, 78, 123
Gold discovered in California, 28
Grand Island, Nebraska Territory, 83 84

Haslam, "Pony Bob," 4, 37-38

Indians, 2-3; Paiute uprising, 35-39; on Plains, 53, 55, 60, 62-63; in Nevada, 65-66, 78-80; attack telegraph line, 80-90

Julesburg, Colorado, 4, 49, 55, 57-58, 84-88, 89-91, 93

Kennan, George, 106-109, 114-117

Lincoln, President Abraham, 74, 96
Little Rock, Arkansas, 31
Los Angeles, California, 31, 153, 169
Louisville, Kentucky, 19, 21

Mark Twain, 34

Marshall, James, 28
Millerism, 15
Mississippi River, 27
Missouri River, 2-3, 23, 26-27, 41, 44, 66
Moore, Jim, 2, 4-10, 34
Mormons, 69
Morse, Samuel F.B., 11-18, 132, 148
Morse Code, 12-13, 19, 22, 141

Nebraska Territory, 2, 23-26, 41, 83
New Orleans, Louisiana, 19, 26
New York, New York, 19, 21, 23, 28, 32, 74-75, 120, 151, 155, 160-163, 167
New York Times, The, 21, 23-26, 67, 74
New York University, 13
Norwalk, Connecticut, 21

O'Brien, Captain Nicklaus J., 85-88, 89-90
Omaha, Nebraska, 41, 44, 49, 55, 64, 66-67, 73, 90, 160

Pacific Telegraph Bill, 44
Placerville, California, 32
Polk, James K., 18
Pony Express, 1-10, 32-34, 35-39, 67, 123
Postal Telegraph, 144-145

Railroad, transcontinental line completed, 118-120
Russian-American Telegraph Company, 100

Sacramento, California, 30, 32, 77
St. Joseph, Missouri, 2, 41, 67
St. Louis, Missouri, 23, 27, 31-32, 66, 142
St. Petersburg, Russia, 104, 107
Salt Lake City, Utah, 4, 30-31, 45, 47, 49-50, 65, 68-69, 71-72, 73, 81-82, 160
San Francisco, California, 2, 29-30, 31, 34,40, 44, 66-67, 74-75, 76-77, 96, 100, 122; earthquake, 144-145, 153, 155, 160-163
Scott's Bluffs, Nebraska Territory, 62
Seward, William H., 112
Siberia, 98-99, 103, 106-110, 112-117
Sibley, Hiram, 47
Stanford, Leland, 118-120

Telegraph, 10; birth and early development, 11-26; "Bee's Grapevine," 32, 36-37, 45; extended to Fort Churchill, Nevada, 39; completed to Salt Lake City, 73; to San Francisco, 74; Indian attack on, 80-88; construction in Canada, 104-105, 111-112; at Promontory Point, Utah, 120; used to control trains, 124; in train robberies, 127-132; transmitting contests, 139-140

United Press, 141
Utah Territory, 30

Vail, Alfred, 16-17
Van Buren, President Martin, 14
Victoria, British Columbia, 100
Virginia City, Nevada, 35-37, 38

Wade, Jeptha H., 73
Ware, Lieutenant Eugene F., 85-90
Washington, D.C., 14, 16, 18, 28, 30, 96, 120, 143, 160-163
Watson, Thomas, 146-147, 160-163
Western Union Telegraph Company, 18-19, 45, 47, 76, 98-99, 111, 116, 120, 124, 133, 138-139, 144-145, 166-167
Wilson, President Woodrow, 160, 163
Wright, Senator Silas, 18

Young, Brigham, 4, 43, 69-71, 73
Yuma, Arizona, 31